Inland Fisheries

Inland Fisheries

V.B. Sakhare
Head
Post Graduate Department of Zoology
Yogeshwari Mahavidyalaya
Ambajogai – 431 517
Maharashtra

2012
DAYA PUBLISHING HOUSE®
New Delhi - 110 002

© 2012 VISHWAS BALASAHEB SAKHARE (1974–)
ISBN 9789351240778

Published by	:	**Daya Publishing House** **A Division of** **Astral International Pvt. Ltd.** **– ISO 9001:2008 Certified Company –** 4760-61/23, Ansari Road, Darya Ganj, New Delhi - 110 002 Phone: 23245578, 23244987 Fax: (011) 23260116 e-mail: dayabooks@vsnl.com website: www.dayabooks.com
Laser Typesetting	:	**Classic Computer Services** Delhi - 110 035
Printed at	:	**Chawla Offset Printers** Delhi - 110 052

PRINTED IN INDIA

Preface

India has made phenomenal progress in inland fisheries sector. Major inland fisheries resources of country comprise (1) capture fisheries of rivers, lakes and estuaries, (2) culture fisheries of the ponds and tanks and, (3) capture-cum-culture fisheries of reservoirs and oxbow lakes. According to various reports, 80-85 per cent of our total inland fish production comes from capture fisheries resources. Development of inland fisheries as a means to increase the availability of proteinous food for the masses and to create employment opportunities for a wide section of our society has long been recognized. The last fifty years have witnessed concerted efforts by the central and state governments to increase productivity from inland water bodies, mainly through the development of a number of scientific management norms for different aquatic biotopes. The indigenous technologies developed at various Research and Development Organizations have played a vital role in ushering in the fish boom in the country.

The book entitled "Inland Fisheries" has been written with the intention to meet the needs of fisheries graduate as well as post-graduate students. This book is also a very important document for the fisheries scientists and professionals engaged in inland fisheries research. The book is divided in thirty chapters. The language of the book is simple and lucid. The subject matter has been explained with the help of finely drawn, simple and well labeled diagrams.

Wherever necessary, tables have been given for clear understanding of the subject. A short reference of the source materials are listed at the end of the each chapter. The subject index is also listed at the end of the book.

I am indebted to my teachers, particularly my esteemed guide Dr.P.K.Joshi, Associate Professor, Dnyanopasak Mahavidyalaya, Parbhani who has been a source of continuous encouragement and inspiration to me.

I wish to express my sincere gratitude to Dr. Sureshji Khursale, President, Yogeshwari Education Society, Ambajogai who has been a source of constant inspiration. I am especially thankful to Dr. R.V. Kirdak, Principal, Yogeshwari Mahavidyalaya, Ambajogai for his encouragement.

I am grateful to my colleagues, students and friends for their valuable co-operation in the preparation of this book.

I am thankful to the researchers, authors of books, publishers and editors whose publications have helped in making out the text.

Without the understanding and encouragement of my wife, I would never have been able to undertake the task of writing the manuscript of this book.

My thanks are due to Mr. Anil Mittal of Daya Publishing House, New Delhi, who not only very well understands but also undertook with pleasure the task of publishing this title. Finally, I will always remain a debtor to all my well-wishers for their blessings, without which this book would not have come into existence.

V.B. Sakhare

Contents

Chapter 1

Riverine Fisheries

India is endowed with a vast expanse of inland waters with an annual runoff of 167.23 million hectare metre. The different river systems in the country, having an estimated linear length of 45,000 km, provide a traditional source of livelihood to thousands of fishermen and the riverine fish contribute significantly to the country's total inland fish production. Out of 14 large rivers, the Ganga and Brahmaptura river systems having a total linear length of 16523 km and a total runoff of 86.96 million hectare meter constitute about 36.72 per cent of the total linear riverine length and approximately 50 per cent of the total run off of all the rivers in the country.

Ganga River System

Out of 14 large rivers, the Ganga river system is the largest river system of India with a combined length of about 12500 kms draining near about one million square kilometer area. Extending between 70° and 80.5°E longitude, it flows through the states of Haryana, Uttar Pradesh, Bihar, West Bengal, Madhya Pradesh, and Rajasthan. Ganga river system accounts for 29 per cent of river length in the country. The river Ganga has a special place in Hindu mythology. It is closely inter-woven with our culture and civilization. Annual rainfall of the Ganga basin ranges from 250 to 4000 mm.

Ganga supports a wealth of fish fauna both in terms of variety and richness. It is the original habitat of Indian major carps and several economically important catfishes. Ganga is also a major source of the riverine spawn, which meets the carp seed requirements of the culture sector to the tune of 30 per cent. The headwaters of Ganga river system in the high altitude of Himalayas are mostly unexplored and, therefore, the commercial fishery is virtually nonexistent due to exploitation problem such as low abundance of commercial fish, inaccessible terrain, poor communication etc. The snow trouts (*Schizothorax sp* and *Schizothoraichthys spp*) and certain catfishes are common in the head-waters of Ganga. Below 1067m, mahseer (*Tor tor* and *Tor putitora*), the katli, *Acrossocheilus hexagonolepis*, certain cyprinoids *L. dero* and *Bagarius bagarius* form the main food fishes. Some fishes such as *Labeo pangusia, L. dero* and catfish *B. bagarius*, adapted to living in rapid waters, descend to warmer waters of the plains.

The stretches of Ganga between Hardwar and Lalgola (1600 km) provides the richest source of freshwater capture fisheries in India. The commercially important fishes and prawns are:

Major Carps
- ☆ *Catla catla*
- ☆ *Labeo rohita*
- ☆ *Labeo calbasu*
- ☆ *Cirrhinus mrigala*

Other Carps
- ☆ *Labeo pangusia*
- ☆ *Labeo bata*
- ☆ *Labeo dero*
- ☆ *Cirrhinus reba*

Clupeoids
- ☆ *Hilsa ilisha*
- ☆ *Setipinna phasa*
- ☆ *Gadusia chapra*

Freshwater Prawns
- ☆ *Macrobrachium rosenbergii*

☆ *M. biramanicum chopral*

☆ *M. lamarrei*

Snakeheads

☆ *Channa marulius*

☆ *Channa orientalis*

☆ *Channa striatus*

☆ *Channa punctatus*

Catfishes

☆ *Mystus aor*

☆ *Mystus seenghala*

☆ *Silonia silondia*

☆ *Wallago attu*

☆ *Pangasius pangasius*

☆ *Bagarius bagarius*

☆ *Rita rita*

☆ *Clupisoma ganua*

☆ *Eutropiichthys vacha*

☆ *Ailia coila*

☆ *Ompak bimaculatus*

☆ *Ompak pabda*

Featherbacks

☆ *Notopterus chitala*

☆ *Notopterus notopterus*

Spiny Eels

☆ *Mastacembelus armatus*

☆ *M. Panchalus*

☆ *Macrognathus aral*

A survey of the stretch revealed an average of 4 fishing villages per 10 km along the bank and density of active fishermen as 4 per village, which gradually increased downstream to 680 indicating higher fishing intensity in the lower reaches. The total catch from market arrivals at selected centers indicates that the commercial fishing is composed of major carps (*C. mrigala, L. rohita, L. calbasu* and *C. catla*) and catfishes (*W. attu, M. aor, M. seenghala, S. silondia, P. pangasius* and *B. bagarius*). The anadromous hilsa contributes significantly to the catch.

The river Ganga and its tributaries have extensive flood plains of Rapti, Ghagra, Gandak, Burhi Gandak, Kosi, Bagmati, Mahananda, and Damodar and Meanders in the middle reaches. With the recession of floods these water bodies are left with young of major carps along with the marshy fauna comprising of *Channa* sp., *H. fossilis, C. batrachus, A. testudineus, Mastacembelus sp,* and *Notopterus sp.*

There are 11 reservoirs constructed in Ganga river system, among which Rihand (Uttar Pradesh), Gandhisagar (Madhya Pradesh) are the largest reservoirs. Farakka barrage is the only hydraulic structure on the main river which obstructs the river flow just before its bifurcation into Bhagirathi and Padma. The barrage was commissioned during year 1975. The barrage serves a vital purpose of flushing the Hooghly during the dry months. Water impounded behind the barrage is regulated to flow into the main river Padma that flows into Bangladesh and to Bhagirathi in India through the feeder canal. Construction of barrage in year 1975 near Farakka (West Bengal) has nearly eliminated the lucrative hilsa fishery above the barrage. The impact of Farakka on hilsa fishery had been severed. When compared to pre Farakka period the average hilsa catch was decreased by 94.61 per cent, 98.12 per cent and 83.05 per cent at Allahabad, Buxar and Bhagalpur respectively.

Investigations conducted by Central Inland Fisheries Research Institute observed a steady decline in fish yield from Ganga river. There is also an alarming downswing in the population of the Indian major carps. The output of major carp spawn has also registered a declining trend in the major spawn collection centers all along the river. The over fishing and some socio-economic factors are responsible for such decline of fish yield. However the environmental changes are also the main factors for the low fish yield. The main sources of pollution in the Ganga are industrial, municipal and agricultural effluents. Survey of Central Pollution Board (1981) showed that 317 major industrial units are operating along the banks of Ganga and its tributaries and only 30 per cent of the units followed some control measures. The industrial effluents cause direct fish kill, destruction of habitat for benthic and planktonic communities and toxicity to organisms. Textiles and other mixed organic wastes causing depletion of dissolved oxygen and high BOD was recorded at Kanpur. The habitat distortion, water quality detoration, alteration of hydrological regime and hydraulic parameters have wide ranging biological impacts including sharp changes in the benthic productivity, fish community structure and fish yield. The average yield of India major carps at Allahabad declined from 21.25 kg/ha during 1961-70 to 9.94 kg/ha in 1981-87. A decline in catfish from 10.11 kg/ha to 5.07 kg/ha was also recorded. Indian shad (Hilsa) was nearly eliminated during 80's when it formed 0.23 kg/ha from 4.86 kg/ha during 1961-70. However the miscellaneous group of fishes registered increase from 10.59 kg/ha during 1961-70 to 13.58 kg/ha during 1981-87. However, at Buxar center an increasing trend was observed except for Hilsa which declined from 83.82 kg/ha to 2.24 kg/ha during 1981-86.

Hook and line was found to be the main gear in the uppermost stretch where no organised fishery was observed. In the middle stretch, gill nets and drag nets are widely operated, besides hook and line, cast net and traps. In the lower stretch of Ganga, various types of gill nets (*chandijal, phansijal*), drag net, seine net (*kochal, konajal, berjal, chatberjal*) dip or lift net (*gara basal; nauka basal*), purse net (*sangla jal*) are the main gears. Other gears *viz.*, falling net (cast net), scoop net, hook and line and traps are also used in this stretch.

The fishing boats used throughout the river are mostly indigenous, nonmechanized and locally built, except for mechanized boat in few stretches. They have been designed to suit local conditions. The simplest and most primitive types of boat used for fishing in the river are the rafts and *dongas*, operated in calm waters. In the larger rivers and estuaries subject to strong current and tidal movements, sturdier plank built boats are used. The boats operated in river Ganga are generally made of wood or tin (CIFRI, 2006). The fishers of river stretch in the state of Bihar generally had small wooden boat, while in the middle stretch from Kanpur to Ghazipur they had small to large boats made of either tin or wood.

Brahmaputra River

The Brahmaputra river system with a combined total length of 4023 km has a catchment area of 51 million hectare, an average and annual runoff of 38 milion hectare meter and an average annual rainfall of 122 cm. The river is known as 'Tsangpo' in the upper reaches, originates in Kubiangiri, east of the mansarover lake in Tibet. It flows eastward at an elevation of 3600m MSL through southern Tibet for a distance of 1600 km parallel to Himalayas. The river then turns abruptly south in a sharp-bend, at an elevation of 2830 m, and flowing in Arunachal Pradesh for a distance of 180 km enters the valley of Assam at Pasighat, at an elevation of 150 m. In the plains it is joined by two other Himalayan rivers, Dibang and Luhit. Flowing further westward through Bangladesh for its last 480 km joining the Ganga at Gualando.

The Brahmaputra on its way to Bangladesh is joined by a large number of tributaries, both from the north as well as south. Characteristically the northern tributaries are large with steep shallow braided channels carrying high silt discharge, whereas, those on the southern bank are deeper with meandering channels, low gradient and lesser silt load.

The Brahmaputra valley is interspread with abandoned of the river, which are subjected to annual inundations. These areas are called 'beels'. These are found in considerable numbers in Lakhimpur, Nowgong, North Kamrup and Goalpara districts.

A survey of fish fauna of this river system in Assam stretch by Motwani *et al.* (1962) revealed the presence of 126 species of fish belonging to 26 families. Out of these 41 represent fisheries of commercial importance. Out of total 42 fish landing centers, Tezpur, Uzam Bazar and Fancy Bazar of Guwahati and Dhubri are the important main centers for recording statistics of fish landings.

Studies conducted by the CIFRI during the years 1972-79 in the three selected stretches revealed a drastic decline in catch at all the centers. However, the data for 1986 and 1987 reflected that the fisheries have shown signs of revival.

Compared to1973 in 1979 it was observed that the major carps have declined from 47.6 t to 8.5 t, catfishes from 58.7 t to 7.3 t and hilsa from 21.63 t to 8.02 t at Guwahati. Similar had been the trend at Tezpur in the upstream and Dhubri in lower reaches. However, landings in 1987 at Guwahati showed a heavy improvement when major carps reached a level of 23.95 t, catfishes 9.72 t and hilsa 9.5 t.

Similarly if the yield per km is also examined it is found that the average catch per km had also declined from 1973 to 1979, being 2.3 to 0.4 t at Guwahati, 0.9 t to 0.27 t at Tezpur and 0.7 t to 0.37 t at Dhubri but improved to 1.6 t during 1987 at Guwahati. Since in the river Brahmaptura the fishing is mostly conducted near the banks, catch per hectare does not carry much meaning.

In the catchment areas of the river system are found thousands of beels affording a very good and lucrative fishery. It is estimated that the beels occupy an area of 50,000 hectare with a production potential of 70,000 tonnes of fishes. Most of these beels are weed infested and as such contain only such fauna, which can withstand the adverse ecological conditions. Mostly it comprises *A. testudineus, C. gachua, C. marulius, C. batrachus, W. attu, N. chitala* etc. However, those with lesser weed infestation contain major carps, *Mystus sp, P. pangasius, S. silondia* etc. also.

Investigations conducted by scientists of CIFRI (1998) revealed the increasing silt load in the river. The tributaries play a key role in maintaining the annual water quality cycle of river Brahmaputra and increase of turbidity over the years clearly reflect the vast quantity of silt discharge in the river by these tributaries every year.

Different factors are responsible for decline in fisheries of Brahmaputra river. Due to accumulation of silt the Brahmaputra river bed is rising alarmingly. The siltation has become such a serious problem that many tributaries often change their course. Heavy siltation and loss of breeding ground is one of the major factors responsible for decline in fisheries of river Brahmaputra and its tributaries. Continuous exploitation of spawners during spawning season and subsequent fishing of juveniles cause maximum damage to the natural recruitment process. Mass destruction of juveniles results in failure of recruitment of the quality fishes and ultimately decline in the overall fishery of the river. Beels are an integral part of the river and during the process of inundation serve as vital spawning ground for major carps because the easily accessible shallow areas of these lakes provide optimum breeding conditions. Indiscriminate and unscientific construction of sluice gates as well as excessive siltation of connecting channels have led to serving link with the main stream. As a result this natural breeding source is gradually being lost.

Table 1.1: Fish Fauna of Brahmaputra River System

Major carps: *Labeo rohita, Catla catla, Cirrhinus mrigala, Labeo calbasu*

Minor carps: *Labeo gonius, Labeo bata, Cirrhinus reba*

Catfishes: *Wallago attu, Channa gachua, Silonia silondia, Pangasius pangasius, Mystus cavasius, Mystus seenghala, Mystus aor, Mystus menoda, Mystus tengra, Mystus vittatus, Rita rita, Clarias batrachus, Heteropneustes fossilis, Bagarius bagarius, Ompak pabda, Ompak bimaculatus*

Hilsa: *Hilsa ilisha*

Miscellaneous: *Settipinna phasa, Gudusia chapra, Notopterus notopterus, Notopterus chitala, Mastacembelus armatus, Mastacembelus aculeatus, Xenentedon cancila, Glossogobius giuris, Amblypharyngodon mola, Nemacheilus spp, Ambassis spp.*

Godavari River

River Godavari is the largest of the peninsular rivers. It starts from Trambekeshwar and towns like Nashik, Kopargaon, Gangakhed, Paithan, and Nanded are situated on the banks of this rivers. After flowing a distance of 1,465 kilometer, it joins the Bay of Bengal in Andhra Pradesh. The main tributaries of the river are Warna, Pravara, Sindhphana, Manjra, Pranhita, Dudhna and Dakshinpurna. Sakhare and Bidkar (2001) conducted the survey on fish fauna, fisheries exploitation and socio-economic conditions of fishermen engaged in fishing of river Godavari at village Raher of Nanded district in Maharashtra. Out of total 21 fish species, *Catla catla, Cirrhina mrigala, Labeo rohita, Labeo calbasu, P. sarana, Wallago attu, Channa sp* have a very high demand. The fish like *Mystus spp* forms the food for poor people. Some fishes like *Chanda nama, Chanda ranga, Glossogobius* etc comparatively of small size, are useful in public health as larvicidal fishes. The crafts used in the Godavari water at Raher are of primitive type, and no modern craft has been introduced for fishing. The gears most commonly used are the gill-nets, cast-nets and hook and lines. The fishermen of the river Godavari at village Raher are extremely poor and generally illterate, most of them belong to caste Bhoi. On the basis of interviews with the fishermen, the per day catch per fishermen was estimated at about 2.5 kg only. Based on the studies by CIFRI, it has been established that the total catch of commercially important species of carps and catfishes markedly declined despite increase in effort and introduction of more efficient gears. Accompanying the decline in total catch were the reductions in catch per unit of effort and average size of the catch, the symptoms of over fishing.

Indus River System

It comprises the main river Indus and its major tributaries like Kabul, Swat, Kurrum, Jhelum, Chenab, Ravi, Beas and Sutlaj. River Indus is the main river of this system with a total length of 2,880 km. It originates in Tibet near the Mansarover lake at an elevation of 5,180 m and passes through inaccessible mountain ranges in Northern Kashmir and emerges out of the hills near Attock (Pakistan). From Attock to its mouth in the Arabian sea, south of Karachi, the Indus transverses the plains of Pakistan for a length of about 1,610 kilometer. The fish fauna of the upper reaches of Indus river has not been fully explored because of its inaccessibility. The main fish species present in the river are *Labeo gonius, Rita buchanani, Exostoma stolicizkae, Trichogaster sp, Nemacheilus spp,* and *Schizopygopsis stoliczkee.*

Krishna River

This river starts from Mahabaleshwar in Maharashtra and after flowing for a distance of 1,400 km from its origin; it joins the Bay of Bengal. It flows through the states of Maharashtra, Karnataka, and Andhra Pradesh. Its length in Maharashtra is 282 kms. Tributaries of river Krishna are the Koina, Warna and the Panchganga. The fishery is mainly supported by *Labeo rohita, Cirrhina cirrhosa, Thynnichthys sp, Aspidoparia spp,* and *Mystus seenghala.*

Cauvery River

It originates from Brahmagiri hills on the western ghats at an elevation of 340m. After flowing 800 kms, it joins the Bay of Bengal in Tamil Nadu. It has the total catchment area of 87900 sq. km. It flows through the states of Kerala, Karnataka and Tamil Nadu. The major fishes present in the river are *Catla catla, Cyprinus carpio, Labeo kontius, L. rohita, L. ariza, Cirrhina cirrhosa, C. mrigala, Mystus aor, M. seenghala, Wallago attu, Silonia silondia, Channa marulius, Notopterus notopterus* and *Tor putitora.*

Narmada River

It is one of the important rivers peninsular India. It originates in the Amarkantak hills at an elevation of 1.057 m in Bilaspur district of Madhya Pradesh. It has a length of 1,312 kms and total catchment area of 98,7963 sq. km. It is mainly fed by seasonal rains. It flows through the states of Madhya Pradesh, Maharashtra and Gujarat. In its last 160 km course, it flows through the plains and fall into the Gulf of Cambay (Arabian sea). The main fisheries of river is supported by *Labeo rohita, L. calbasu, L. fimbratus, L. bata, Tor tor, Cirrhina mrigala, C. reba, Puntius sarana, Mystus seenghala, M. cavasius, Clupisoma garua, Ompak bimaculatus* and *Wallago attu*. Other miscellaneous fishes are *Channa spp, Mastacembelus spp* and *Notopterus notopterus*.

Mahanadi River

The Mahanadi, rising in Sihawa hills to the extreme south-west of Raipur district of Madhya Pradesh, drains an area of 141600 sq. km. of which 53 per cent is in Madhya Pradesh, 46.4 per cent in Orissa,0.5 per cent in Bihar and 0.1 per cent in Maharashtra. The total length of the river is 857 km and has a maximum discharge of 44740 cubic m. sec. and an annual flow of 66640 million cubic meter. The river is totally rain fed and hence nearly 80 per cent of the run-off occurs during the monsoon season. The fishery of river comprise the carps (*Catla Catla, Labeo rohita, L. calbasu, L. gonius, L. fimbratus, L. bata, Cirrhina mrigala, C. reba, Tor mossal*), catfishes (*Wallago attu, Ompak bimaculatus, O. pabda, Silonia silodia, Mystus spp, Pangasius pangasius, Bagarius bagarius, Rita chrysea*, and *Eutropiichthys vacha*), live fishes (*Channa spp, Heteropneustes fossilis*, and *Clarias batrachus*), featherbacks (*Notopterus sp*), prawns (*Macrobrachium malcolmsonii*) and miscellaneous species (*Puntius sp, Rohtee cotio, Sciaenid spp, Mastacembelus armatus, Nandus nandus, Glossogobius giuris, Gadusia chapra, Esomus danrica, Rasbora daniconius, Amblypharyngdon mola* etc.).

Scientists of CIFRI (1997) conducted a survey of river Mahanadi during 1995-96 to assess the status of fisheries and environment. The survey revealed existence of 78, 24 and 110 fish species in the upper, middle and lower stretches respectively, indicating a rosy picture of fish biodiversity. Fish landings are at a optimal level. Average upstream catch (25-100 kg/day/centre) comprises mainly catfishes (40 per cent) and minnows (35 per cent). Catch at Sonepur (248 kg/day) is higher than the average catch of the middle stretch (36 to 122 kg/day/centre), because carps are the prime contributors. The fish yields of the lower zone are 50-250 kg/day/site at freshwater stretch up to Mundh barrage, 400-600 kg/day in the stretch between

Mundah barrage and Cuttack barrage and 10-750 kg/ha/site in the estuarine stretch. Due to availability of coastal fishes in the estuarine zone the yield at Paradip touches nearly 18-32 t/day in winter months.

Biotic ecosystem of the river is moderately congenial. The ranges of physico-chemical parameters like water temperature, dissolved oxygen, pH, nitrates, phosphates etc are within the optimum range but the salinity in the estuary is quite low. Specific conductivities at places are quite high.

Medium and Minor Rivers of India

India is having 44 medium rivers with a total drainage area of 0.24 million sq. km. Nine of these rivers flows through more than one state and are considered as interstate rivers. Seventeen medium rivers flow towards west into the Arabian sea, and twenty-three towards the Bay of Bengal while four rivers in Mizoram and Manipur flow into Bangladesh and Burma. Minor rivers has a drainage area of less than 2000 sq. km. They are numerous and are essentially small streams, flowing from the western and eastern ghats into the sea. Their total catchment area is 0.2 million sq. km.

References

Chandra, Ravish. 1989. Riverine fisheries resources of the Ganga and the Brahmaputra. In: Conservation and management of Inland capture fisheries resources of India (Eds. Arun G. Jhingran and V. V. Sugunan), p. 52-60, Inland Fisheries Society of India.

Central Inland Fisheries Research Institute (CIFRI), Barrackpore. 1997. Rapid survey of River Mahanadi. The Inland Fisheries News (Newsletter of the Central Inland Capture Fisheries Research Institute. 2(2): 1-2.

Central Inland Fisheries Research Institute (CIFRI), Barrackpore. 1998. Decline in fisheries of River Brahmaputra: An ecological study. The Inland Fisheries News (Newsletter of the Central Inland Capture Fisheries Research Institute. 3 (2): 1-2.

Central Inland Fisheries Research Institute (CIFRI), Barrackpore. 2006. Report of ICAR AP Cess Fund Scheme"Techno-Socio-economic status of Fishers of River Ganga" 80 p.

Central Inland Fisheries Research Institute (CIFRI), Barrackpore. 2008. Current Status of River Ganges, p. 34.

Sakhare, V. B. and Bidkar, A. R. 2001. Fisheries status of river Godavari at village Raher of Nanded district in Maharashtra, India. *Ecol. Env. and Cons.* Vol. 7(2), pp. 151-153.

Chapter 2
Reservoir Fisheries

The hydraulic structure across a river to store water on its up-stream side, which is an impervious or fairly impervious barrier across a natural stream so that a reservoir is formed. This water is then utilized as and when it is needed. Due to construction of reservoir, water level in that river at its upstream side is very much increased, and a large area may be submerged depending upon the water spread of the reservoir so formed. Due to increased demands for reliable supplies of electric power, irrigation, and drinking water, the number of new hydropower reservoirs are increasing dramatically, especially in Asia.

It has been estimated that annual inland fish production in Asia is 5.5 million tons, comprising 57 per cent of the world's inland fish production (De Silva, 1988, 1992). However fish yields from Asian reservoirs comprise just 0.5 million tons of this 5.5 million tons (De Silva, 1988). De Silva (1988) estimated fish production from Asian reservoirs at only 20 kg/ha/year with a wide variability in production that was not always related to the size of the reservoir. Costa- Pierce and Soemarwoto (1987) estimated an average percentage increase in reservoir area in 15 Asian nations from 1987 to 2000 would be 511 per cent ranging from 50 per cent (Singapore) to 9,900 per cent (Laos). By 2000 it is predicted that the collective water surface in reservoirs (20.3 million ha) will exceed the surface area of Asia's natural waters (18.5 million ha). Clearly if the huge expanse of under utilized water areas locked behind Asia's dams could be utilized for increased fish production, thousands of tons of new aquatic protein could enter the Asian markets (Costa-Pierce, 1998). Production of new aquatic protein is especially urgent in Asia, a region where fish is the most important source of animal protein. In addition, there is a need to create thousands of new rural jobs due to population growth is evident and to final innovative ways to stem the rapid rural to urban migration. It is argued that expansion of aquatic food production in Asian reservoirs could assist in mitigating Asia's growing food and population crises (Brown, 1997).

Exposed to the warm tropical climate, these water bodies in our country are extremely productive and they harbour an enviable spectrum of fish genetic resources. Optimum utilization of these resources can lead to a man-fold increase in inland fish production, earning the country a place among the top inland fish producing nations of the world. As a result of the construction of dams under various multipurpose river valley projects, a number of fresh water impoundments have come into existence. These impoundments are called reservoirs or dam-lakes which represent the complex environmental and biological systems. Within a reservoir there is a great diversity of interactive components ranging widely in morphometrical, operational and biological characteristics.

Reservoirs are generally classified as small (<1000 ha), medium (1000-5000 ha) and large (>5000 ha), especially in the records of the Government of India (Srivastava *et al.*, 1985). Jhingran and Sugunan (1990) classified the reservoirs as large (>1000 ha), medium (500-1000 ha) and small (< 500).

Reservoirs form the largest inland fisheries resources in terms of resource size with 56 large (>50000 ha), 180 medium (1000-5000 ha) and 19134 small reservoirs covering a water area of 1.14 m ha,0.527 ha and 1.2 m ha respectively with substantial areas added year after year due to construction of new impoundments created through erection of dams over rivers, streams or any other water course (Chakraborty *et al.*, 2011).

The total water spread area of Indian reservoirs is about 3 million hectares, which forms about 50 per cent of the total reservoir area in

South East Asia (Shetty 1990). Indian Institute of Management, Ahmedabad (Srivastava *et al;* 1985) listed 975 reservoirs in India, in size range of 1000 ha to 10,000 ha covering an area of 1.7 million hectares. According to Sugunan (1995), the present total area under the reservoirs in our country is 3.1 millon hectares (Table 2.1).

Table 2.1: Distribution of Small, Medium and Large Reservoirs in India

	Small Reservoirs	Medium Reservoirs	Large Reservoirs	Total
Number	19134	180	56	19370
Area (ha)	1485557	527541	1140268	3153366

Data pulled from Sugunan (1995): FAO Report.

According to Dixitulu (1999) there are 19, 370 reservoirs in the country with an extent of 3,153,366 ha (consisting of 19,134 nos of small ones with an extent of 1,485,557 ha; 180 nos of medium ones with 575,541 ha and 56 nos of large ones with an extent of 1,140,268 ha) Kaur and Dhawan (1997) mentioned total number of 550 reservoirs with a total water spread area of about 2 million hectares. Sugunan (1995) estimated the present annual yield per ha from small reservoirs (less than 1000 ha each) at an average of 49.90 kg, from medium reservoirs at 12.30 kg and from large reservoirs at 11.43 kg. He gives the pooled production for all the three categories of reservoirs as 20.13 kg/ha. Based on these estimates the present total annual production from Indian reservoirs has been computed at 93, 650 t, against the potential of 245, 134 t (from 3,153,366 ha); or in other words, the average per hectare potential production has been visualized is around 80 kg/ha/yr (Dixitulu, 1999). He further mentioned that going by this formulation, 245,134 t (from 3, 153,366 ha). In other words, the average per ha potential production the export has been visualized is around 80 kg. Going by this formulation 245, 134 t should generate gross returns of Rs. 735 crores at an average realization of Rs. 30/- per kg as pooled value, realized out of domestic sales and exports through centralized storage plants and the connected cold chain and domestic marketing system. The present Indian reservoir fish production is frustratingly low; the various estimates of national average ranging from 6 to14 kg/ha/yr (Jhingran, 1985). Indian reservoirs, being in the tropics, have high primary productivity and have the capacity to produce more fish than their present low Indian average of 20-25 kg/ha/yr in large reservoirs and 49.5 kg/ha/yr in minor reservoirs (Piska, 1999). It is also estimated that the reservoir fisheries development has the potential to generate additional national income to the tune of Rs. 100 crores per year and can provide employment to millions of fishermen and workers in the ancillary industries (Srivastava and Reddy, 1983). Dwivedi (1980) states that if the optimum, productivity level is achieved from present water area of reservoirs alone can provide more than 50,000 tonnes of fish and can provide employment to nearly 72,000 fishermen. Khan *et al.* (1991) mentions 3 million hectare reservoir area in our country offer tremendous scope for augmenting fish production; and reservoir fishery is also important from sociological point of view with respect to providing employment to about 2 million people.

Unfortunately, majority of reservoirs are not being scientifically managed for fisheries. Only a handful of them have so far been harnessed along scientific lines, while the others are either half-heartedly managed or not managed at all. As compared to several developed countries, the per hectare fish production in Indian reservoirs is very poor, being only about 20-25 kg per year as against 88 kg in the USSR (Jhingran, 1985), 100 kg in Sri Lanka and 64.5 kg in large reservoirs of Thailand (Bhukaswan, 1977). Dixitulu (1999) mentioned that in case of reservoir fisheries, however, the developmental process has been tardy, remaining by and large, at the level as at the time the British withdrew, with the exception of a few reservoirs. It is to the credit of the British that, at quite a few reservoirs, they set up farms and also undertake stocking of the reservoirs. The process no doubt continued after independence, with the production being conspicuous in a very few reservoirs, below average in some, and low to very low in others. One of the main factors responsible for low fish yield from Indian reservoirs is the unscientific management practices, which stem from the inadequate knowledge of the ecology and production functions of this biotope. Apart from the urgent need to take up all old reservoirs for proper appraisal of their present status and for taking up rational management and conservation measures, it is highly imperative to take up fisheries development work in new reservoirs right from the initial stage itself. Lot of progress has been made in several developed countries in deep water fishing in reservoirs and management of

Figure 2.1: A Panoramic View of Ujani Reservoir of Maharashtra

their fisheries. The neglect of this vast potentially rich fishery resource is all the more appalling, in view of inadequate fish production in the country even to meet the protein demands of our exploding populations, let alone the export demands.

Reservoir Fisheries in Different States

Tamil Nadu

Except in some states like Tamil Nadu it is difficult to get data on even total production of a state. In southern region of country, in Tamil Nadu the total annual fish production potential for reservoirs is about 2,000 tonnes, which indicates per unit area production of 40 kg (Sanskarasubbaiyan and Menon 1984). However, Sreenivasan (1998) mentions 52,000 hectare of reservoir area. Sugunan (1995) estimated 52 small, 8 medium and 2 large reservoirs with 48.50, 13.74 and 12.66 kg/ha/yr fish yield respectively. CIFRI (1998) surveyed Amaravathy, Palar- pornthalar, Uppar, Pillor, Gunderipallam and Varattupallam reservoirs of Tamil Nadu and revealed that the fish yield in Varattupallam was highest and lowest in Amaravathy. Taking into consideration the area of reservoirs and stocking density, it was concluded that the fishery management in Palar- Poranthalar is best followed by Amaravathy and uppar, while it was poor in Pillar. The actual fish yields obtained from the reservoirs are less compared to the production potentials estimated through primary production studies. This gives scope for enhancement of fish production from these reservoirs through judicious stocking and exploitation.

Kerala

Being essentially a marine fisheries state, inland fisheries till recently did not receive much attention. There are 29,660 ha of reservoirs with a fish yield less than 15 kg/ha; mostly Tilapia (Sreenivasan, 1998). According to Sugunan (1995), state is having 7 small and 2 medium reservoirs with fish production of 53.50 and 4.80 kg/ha respectively. Further Sugunan also mentions, there are total 30 reservoirs with water-spread area of 29,635 hectare. Mohan(2001) highlighted the economic and social conditions of fishermen of Malapuzha reservoir and concluded that 79.17 per cent or respondents were literate.

Karnataka

Karanataka is an important maritime state, the marine fisheries in state is more dominant than inland fisheries, where a major finance has been expended for the development of marine fisheries. State is having 60 reservoirs with an area of 2.20 lakh hectares (Annual Report, Dept. of Fisheries 1994-95). However, IIMA (1985) mentions presence of 28 reservoirs with total water spread area of 1,51,624 hectares, the average fish production from which is about 23-35 kg/ha/yr. Sugunan (1995) mentions the total area under reservoir in the state is 4,37,291 hectares comprising 4,679 reservoirs of different categories. Sukumaram (2001) studied status of plankton from Ghataprabha, Malaprabha, Manchenebele, Nugu, Hemavathy, Kabini, Linganamakki, Harangi, V. V. Sagar and Narayanpur reservoirs of the state and concluded that all the reservoirs barring Ghataprabha are free from stress.

Andhra Pradesh

It has 2.143 lakh hectares of reservoirs of all sizes (Sreenivasan, 1998). According to Sugunan (1995), the state is having 37 small, 29 medium and 3 large reservoirs having 188, 22 and 16.80 kg/ha fish yield respectively. IIMA (1985) estimates the average reservoir fish production of 37.43 kg. CIFRI (1998) also conducted the survey of nine reservoirs of Andhra Pradesh and found that most of the reservoirs have the necessary infrastructure for raising the seed and exploitation of the fishery and recommended that the co-operative societies need to be activated and given the responsibility of management. There is an immense scope to introduce cage culture systems in most of the reservoirs. This will enhance the yield from reservoirs and also provide additional employment.

Rajasthan

It is with 78 small, 17 medium and 2 large reservoirs with a fish yield about 46.43, 24.47, and 5.30 kg/ha/yr respectively (Sugunan, 1995). According to IIMA (1985), the total number of reservoirs in the state is 665.

Uttar Pradesh

In Uttar Pradesh all the reservoirs are multipurpose. According to Directorate of fisheries, Uttar Pradesh, the state is having 2.83 and 1.62-lakh hectares area under large and small reservoirs respectively. Sugunan (1995) estimates 31 small, 13 medium sized and one large reservoir in the state having fish yield of 14.60, 7.17 and 1.07 kg/ha/yr respectively. Punjab has less potential for reservoir fisheries. There are four dams. The naturally available fish species are the major carps, common carps, *Tor putitora, Labeo dero, Labeo bata, Mystus seenghala, Synothorax* etc. (IIMA, 1985).

Orissa

In Orissa the total reservoir area is 99,061 ha (IIMA, 1985). Sreenivasan (1998) estimated about 1,19,403 ha large, 9,246 ha medium and 65047 ha small sized reservoirs. According to Sugunan (1995), the state is having a total of 1442 reservoirs with an area of 1,98,198 ha.

West Bengal

It is an important maritime state on the east coast of India having only 7 reservoirs, out of which two are major ones (IIMA, 1985). According to Sugunan (1995), the state is having six reservoirs with combined water spread area of 15,732 hectares. No correct data on fish yield, number of reservoirs and stocking is available as the pond culture practices are more dominant than the reservoir fisheries. In Bihar, total reservoir area is about 50,000 hectares. There are 350 fishermen's co-operative societies in the state with total membership of 27,142, which are under control of state co-operative department. The functioning of these societies are not satisfactory and are dominated by influential persons who exploit the members.

Madhya Pradesh

It is purely an inland fish producing state. According to Sugunan (1995), the total number of reservoirs in the state is 32 with water-spread area of 460384 hectares. IIMA (1985) reported the large and small reservoirs with a total water spread area of 2.5 lakh hectares. The fish production for the period of 1974-75 to 1979-80 from selected reservoirs was varied from 0.20 to 87 kg/ha/yr (IIMA, 1985). According to Sreenivasan (1998), state is having 1.17-lakh hectare of reservoirs yielding an average fish catch of 26.46 kg/ha. George (2001) studied Gandhisagar, Bargi, Tawa, Barna and Halali reservoirs to compare the efficiency of the management measures followed. In the present scenario, ownership of these water bodies by apex federations of primary fishermen cooperatives seems to be the best option for their management for their management and sustainable exploitation.

Gujarat

The Gujarat has about 0.14 million ha. area under reservoirs (IIMA, 1985). According to State Fisheries Department of Gujarat, the total area under reservoirs in the state is 286230 hectares and more than 95 per cent (in the number)of these manmade lakes belong to the small category, although they form only 29 per cent of the total area. Anon (1980, 1983, 1984b) furnished the information on few reservoirs in the state.

Maharashtra

It is one of the largest states in the country in population and geographical area having a number of rivers like the Godavari, the Bhima, the Krishna, the Narmada, the Tapti and other several rivers and their tributaries having a total of 1600 km of river length. Sugunan (1995) mentions the total reservoir area in the state is 2,73,750 ha. However, according to Sreenivasan (1998), Maharashtra is endowed with 1,79,430 ha. of reservoir area and the state produced 7.83 kg/ha fish from its reservoirs. However IIMA (1985) worked out 1,05,202 ha. reservoir area comprising 72 reservoirs. Pathak

(1990) is of the view that, the area under reservoirs in the state of 20 ha and above estimated to about 2,36,157 hectares. No limnological studies or survey of fish fauna has been carried out so as to know the productivity of water and fish faunal diversity. However, Sugunan (1995), Sreenivasan (1991), Valsangkar (1980 & 93), Sakhare (1999, 2001, 2003, and 2005), Sakhare and Joshi (2002, 2003, and 2004), Sakhare and Jetithor (2011), IIMA (1985), and Desai (1980) documented information on fisheries and/or ecology of some of the reservoirs in the state.

Crafts and Gears Used for Fishing

The success of reservoir fisheries management depends on the quality of fish seed, proper stocking, exploitation of fish with appropriate crafts and gears, providing ice and cold storage facilities for better marketing of fish. Crafts used in the Indian reservoirs are of primitive type, and no modern crafts have been introduced for fishing. The craft called 'Nav' is common in reservoirs of Maharashtra. This craft is nothing but a platform of 6x3 feet size with a depth of 4-10 inches is prepared from the thermocoel and covered by a plastic covering. The cost of one craft varies from 400-600 in the local markets. Generally the motorboats are not used because of heavy expenditure and other mechanical problems. The cooperative societies engaged in fishing are not in a position to provide loans or subsidy for the purchase of wooden boats/motorboats; because the members have not returned the loans already taken. Devi (1997) accounted traditional catamaran improved with thermocoel and coracles from Ibrahimbagh and Shathamraj reservoirs of Hyderabad. There are so many drawbacks of such type of crafts such as :1) They are not suitable to carry large nets, having heavy weight.2)When fishermen get large quantities of fish catches, it become difficult to carry them to the landing centers.

Gears used for the fishery are mostly the surface gill-nets. However, cast-nets and hook and lines are also used. The cost of net varies according to the quality of nylon used for their fabrication. Good quality nets are prepared from the Garware nylon, which is considered good quality nylon yarn. Khan *et al.* (1991) mentioned that the gill-nets used in almost all reservoirs require improvements in respect of design, rigging and other parameters like mesh size, and twin size in order to increase its efficiency.

Figure 2.2: Fishing in Progress

Seed Stocking

According to Raghavachari (1983) stocking is not being done systematically in Indian reservoirs. There is a shortage of seeds and fingerlings. The size of fingerlings stocked has a direct bearing on the production at the later stage. Many reservoirs do not have a fish seed farm at the site. Even if there is one, the seed farm is unable to meet the needs of the reservoir. Raghavacharis and Surychandra Rao (1984) and Srivastava (1985) recommended the following stocking rates:

☆ Large reservoirs (above 5000 ha)–200 fingerings/ha

☆ Medium reservoirs (1000-5000 ha)–400 fingerlings/ha

☆ Small reservoirs (less than 1000 ha)–1000 fingerlings/ha.

Raghavacharis and Suryachandra Rao (1984) further mentions that the stocking should be done in one size in the form of advanced fingerlings with size of 12.5 cm. Paul and Sugunan (1990) reported the stocking programme in reservoirs had been generally devoid of any ration are behind them and consequently, no desirable impact was visible in most of the cases. They also stated that in many reservoirs stocking programme on scientific basis were impaired due to lack of seed availability. They suggested that the matters regarding selection of stocking materials, size of fingerlings and the stocking rate should be decided after taking into account the relevant ecological data. Selvaraj and Murugesan (1990) reported the factors responsible for low yield in Aliyar reservoir of Tamil Nadu, one of the main factor was stocking. The size of fry stocked was usually so small that they fall an easy prey to the predatory fishes. Even when the seed of major carps was produced in the fish farm through hypophysation, the size of seed stocked remained very small leading to their high mortality and subsequent poor yield and low contribution by the major carps in the landings. Sinha (1998) recommended a rate of 250 fingerlings per ha for reservoirs without catfishes and 600 fingerlings (6" in length) with rich catfish population. But, for new reservoirs, the stocking rate should be at a higher level (1000 kg/ha). Irrational stocking of fingerlings has a deleterious effect on reservoir fisheries. However, a rational approach for formulating stocking policy is through estimation of potential fish yield of the reservoir and adjustment of stocking rate in such a manner as to obtain the yield close to the productivity potential. Stocking of economically important, fast growing fishes from outside is aimed at colonizing all the diverse niche of the biotope for harvesting maximum sustainable crop from them. Central Inland Fisheries Research Institute (1997) under the All India co-coordinated Projects on Ecology and Fisheries of Freshwater Reservoirs carried out investigations on Bhatghar reservoir of Pune district of Maharashtra. The investigation recommended the need for regular stocking of major carp fingerlings. Sakhare (2003) reported autostocking in Yeldari reservoir of Maharshtra. The similar type of observations were also made by Devi (1997) in Ibrahimbagh and Shathamraj reservoirs of Hyderabad.

According to Sreenivasan (1989) when existing species are inferior, it would be necessary to add superior varieties. Species esteemed by consumers (taste, texture, flavour) have to be introduced. For Indian reservoirs, the following species are suitable- *Catla, Rohu, Mrigal, Labeo calbasu. Cirrhinus cirrhosa, Mahseers, Labeo bata,* and in some cases *L. fimbriatus.* Among exotics the common carp (*Cyprinus carpio*), Bangkok variety is suitable only for certain reservoirs. The Chinese carps, grass carp (*Ctenopharyngodon idella*) and silver carp (*Hypophthalmicthys molitrix*) have not been viewed with favour and have raised a dust of controversy. The example of role of silver carp in Gobindsagar is cited to prove the adverse effect of this species in reservoirs. These species found in Gobindsagar conditions similar to its native China and spawned prolifically and continuously. The same situation is not likely to occur in reservoirs warmer than this and further down south. *Tilapia mossambica* is not a recommended species for reservoirs with good carp fishery. Pathak (1990) suggested fingerlings of 100-150 mm size of commercially important species (*Catla, Rohu, Mrigal* and Common carp) need to be stocked in reservoirs, which would withstand the pressure of predation to some extent. He further recommended to stock large, medium and small sized reservoirs at the rate of 500, 1000 and 1500 fingerlings/ha respectively. He also experienced that whichever reservoir was stocked properly in the past, they gave a very good fish production.

Fish Fauna and Fish Production

Depending on the available literature, the fish production from small, medium and large sized reservoirs in India varies from 0.52 to 912.7 kg/ha/yr. The highest rate of fish production was recorded from small reservoirs and lowest from large sized reservoirs. Sugunan and Yadava (1992) Sakhare (1999, 2001, & 2003), Sakhare and Joshi (2002 &2003), Sakhare and Jetithor (2011), Valsangkar (1993), Singh (2001), Piska *et al.* (2000) and many other workers reported the fish fauna of various Indian reservoirs. The common fishes recorded from Indian reservoirs are listed in Table 10.2. The checklist is prepared with the help of research papers published by various scientists and institutions in country.

Table 2.2: Fish Fauna of Indian Reservoirs

Order: Clupiformes–*Notopterus notopterus, N. chitala*

Order: Perciformes–*Chanda nama, C. ranga*

Order: Cypriniformes–*Salmostoma boepia, S. untrahi, Rasbora elanga, Rasbora danoconous, Danio equipinnatuus, D. devario, Catla catla, Labeo rohita, L. calbasu, L. fimbratus, L. boggut, L. porcellus, L. sindensis, Cirrhina mrigala, Cirrhina fulungee, Cyprinus carpio, Garra lamta, Hypophthalmicthys molitrix, Osteobrama vigorsii, O.neilli, O. cotio cotio, Rohtee ogilbii, Tor khudree, T. mussal, Puntius sophore, P. kolus, P. ticto ticto, P. sarana, P. jerdoni, P. amphibious, P. subnastutus, Nemachilus evazardi, Nemachilus denisonii, Nimachilus rueprel.*

Order: Siluriformes–*Ompak bimaculatus, O. pabda, Silonia silondia, Heteropnestus fossilis, Wallago attu, Pseudotropius etherionoides, Mystus seenghala, M. cavasisus, M. vittatus, M. gulio, M. aor, M. tengra, Xenentodon cancilla, Rhinomugil corsula, Channa marulius, C. gachua, C. striatus, Bagarius bgarius, Pangasius pangasius, Clarias batrachus, Glossogobius giuris giuris*

Declaration of Closed Season and Poaching

Raha and Sarkar (1980) reported the declaration of closed season during monsoon season (from April to June), when no fishing is allowed in the reservoir. Anybody found fishing in the reservoir during the closed season or without licence during the fishing season is liable to be prosecuted before the court under Indian Fisheries Act.

Poaching is also a notable thing in many reservoirs as there is no sufficient staff to prevent it. George (2001) also mentioned poaching as a major problem in Gobindsagar, Bargi and Tawa reservoirs of Madhaya Pradesh. It is due to the lack of manpower, heavy costs, vastness of the area, hostile terrain and absence of vehicles. According to Raghavacharis and Rao (1984) in some Indian reservoirs poaching yields greater production than the legal fishing and it is difficult to check poaching because of vastness of the reservoir and the terrain around the reservoirs. Negligible poaching was observed in Gobindsagar and Pong dam; they are also of the view that due to high cost, an official agency alone cannot check the poaching. Sugunan and Yadava (1991) mentioned necessity of boat which will serve to transport fish from gill net points to the landing center and the same boat can be used for patrolling the Nongmahir reservoir (Meghalaya) to ward off poachers and to check undesired fishing activities.

Deforestation of Standing Trees

Deforestation of standing trees in many reservoirs has always remained neglected. Ringangaonkar and Bhrushundi (1980) described the manual saw to cut down submerged standing trees in Nalganga reservoir. Trees from the area which were exposed during summer months when water level of reservoir goes down. In case of trees in deeper waters where saw cannot reach, the branches are cut down. Applying forces for cutting by a one person also cut standing trees in shallow water. But to cut the tree under water the saw should reach the bottom of the tree-ordinary saw operated by two persons was not sufficiently long to reach the bottom. The saw length was increased by fixing long handles According to Jhingran (1985) the problem of clearance of submerged obstruction was squarely tackled in a number of small reservoirs constructed in Madhya Pradesh, Chillar and Benisagar being well known examples where from the standing on reservoir bed were cut and the wood auctioned. Even

Table 2.3: Few Important Reservoirs in India

Sl.No.	Reservoir	State	Area (ha)	Year of Construction	River
1.	Mettur	Tamil Nadu	15346	1934	Cauvery
2.	Bhavanisagar	Tamil Nadu	7886	1953	Bhavani
3.	Sathanur	Tamil Nadu	2010	1957	Ponniar
4.	Stanley	Tamil Nadu	15346	1939	Cauvery
5.	Malampuzha	Kerala	2313	—	Bharatapuzha
6.	Idukki	Kerala	6160	1973	Periyar
7.	Parappar	Kerala	2590	1987	Kallada
8.	Chulliar	Kerala	165	1970	Chulliar
9.	Tungabhadra	Karnataka	37814	—	Tungabhadra
10.	Markonahalli	Karnataka	1336	1939	Shimsha
11.	Hemavathy	Karnataka	9162	1981	Hemavathy
12.	Vanivilas Sagar	Karnataka	8640	1901	Vedavati
13.	Krishnarajsagar	Karnataka	13200	1932	Cauvery
14.	Nagargunasagar	Andhra Pradesh	28474	1967	Krishna
15.	Yerrakalava	Andhra Pradesh	1550	1988	Yerrakalava & Jalleru
16.	Hussainsagar	Andhra Pradesh	446	1850	Kukatpally
17.	Yeldari	Maharashtra	9472	1962	Purna
18.	Nathsagar	Maharashtra	39777	1975	Godavari
19.	Ujini	Maharashtra	29000	1978	Bhima
20.	Siddeshwar	Maharashtra	3360	1962	Purna
21.	Palas Nilegaon	Maharashtra	206	1979	Bori
22.	Bori	Maharashtra	746	1966	Bori
23.	Manjara	Maharashtra	4590	1981	Manjara
24.	Bhatghar	Maharashtra	2800	1928	Yelwandi
25.	Gangapur	Maharashtra	2400	1973	Godavari
26.	Shivajisagar	Maharashtra	11535	1961	Koyana
27.	Girna	Maharashtra	5420	1970	Girna & Panzan
28.	Jawalgaon	Maharashtra	858	1977	Nagzari
29.	Hingni pangaon	Maharashtra	1006	1976	Bhogawati

Contd...

Table 2.3–*Contd...*

Sl.No.	Reservoir	State	Area (ha)	Year of Construction	River
30.	Ekruk	Maharashtra	1842	1871	Aghila
31.	Khadakwasla	Maharashtra	1472	1967	Mutha
32.	Mula	Maharashtra	5324	1978	Mula
33.	Bhandardara	Maharashtra	1822	1926	Pravara
34.	Dhom	Maharashtra	2308	1977	Krishna
35.	Manar	Maharashtra	2460	1964	Manar
36.	Kodar	Madhya Pradesh	3583	1981	Kodarnala
37.	Gandhisagar	Madhya Pradesh	66000	1960	Chambal
38.	Ravishankar sagar	Madhya Pradesh	9540	1978	Mahanadi
39.	Govindgarh	Madhya Pradesh	307	1916	Bichia
40.	Tawa	Madhya Pradesh	20055	—	—
41.	Bargi	Madhya Pradesh	27296	—	—
42.	Halali	Madhya Pradesh	7712	—	—
43.	Hirakud	Orissa	71963	—	Mahanadi
44.	Rangali	Orissa	37840	—	Brahmani
45.	Getalsud	Bihar	3459	1971	Subarnarekha
46.	Konar	Bihar	2590	1955	Konar
47.	Maithon	Bihar	11490	1957	—
48.	Panchet	Bihar	7511	1958	Damodar
49.	Badua	Bihar	880	1963	Badua
50.	Nalkari	Bihar	991	1967	Nalkari
51.	Rihand	Uttar Pradesh	46538	1962	Rihand
52.	Baghla	Uttar Pradesh	250	1952	Barica
53.	Nanaksagar	Uttar Pradesh	4662	1957	Chuka
54.	Gobindsagar	Himachal Pradesh	16867	1959	Sutlej
55.	Pong	Himachal Pradesh	24629	1974	Beas
56.	Jaismand	Rajasthan	7286	1730	—
57.	Ramgarh	Rajasthan	1260	1903	Banganga
58.	Chhparwara	Rajasthan	200	1894	Mavshi
59.	Mahibajaj	Rajasthan	13500	—	—

Contd...

Table 2.3–*Contd...*

Sl.No.	Reservoir	State	Area (ha)	Year of Construction	River
60.	Kadana	Rajasthan	9000	—	—
61.	Gudha	Rajasthan	1928	—	Mej
62.	Keetham	Rajasthan	306	1925	Jamuna
63.	Vallabhsagar	Gujarat	52000	1971	Tapti
64.	Kadana	Gujarat	16600	1979	—
65.	Panam	Gujarat	8980	1977	—
66.	Gumti	Tripura	4500	1976	Gumti
67.	Nongmahir	Meghalaya	70	1979	Umiam
68.	Khandong	Assam & Meghalaya	12950	1986	Kapili
69.	Kyrdemkulai	Meghalaya	90	—	Umiam
70.	Umrong	Assam	5550	1986	Kapili

after the reservoir is filled it is possible to clear the obstructions at least in the reservoir margins but at that stage the work can be taken up only in the summer months when the water level goes down and the shore areas are exposed. He further mentioned that the presence of aquatic vegetation not only hampers fishing but also suppresses growth of benthic fauna as it occupies the productive zones. A machine has been developed at the Central Institute of Fisheries Technology for eradication of submerged weeds, which consists of a rotating fork arrangement driven by an engine, the rake and engine having been installed in a self-propelled pantoon. The cost of clearing works out to be Rs. 100/- per hectare. A clearance of about 80 per cent is obtained. About 1 to 1.25 ha per day of 8 hours operation can be cleared.

References

Anon, 1980. Proceedings of the 6th workshop of All India Coordinated Project on Ecology and Fisheries of Freshwater Reservoirs, Simla 25-26 November 1980, Central Inland Fisheries Research Institute, Barrackpore, West Bengal, pp 121.

Anon, 1983. Proceedings of the 7th workshop of All India Coordinated Project on Ecology and Fisheries of Freshwater Reservoirs, Barrackpore, 9-10 March 1983, Central Inland Fisheries Research Institute, Barrackpore, West Bengal, pp 148.

Anon, 1984 b. New low cost coracle for fishermen. CIFRI Newsletter, 7(1&2): 3-4.

Brown, L. 1997. Facing the prospect of food scarcity. In L. Brown (Ed.) State of the world. Washington, Dc: Worldwatch. pp 23-41.

Chakraborty, S. K., Sreekanth, G. B., Jaiswar, A., and Ambarish P. Gop. 2011. Inland fisheries resources of India: A critical overview. *Fishing Chimes.* 31 (1): 32-37.

CIFRI. 1997. Ecology and Fisheries of Bhatghar reservoir. Golden jubilee spl. Bulletin No. 73.

CIFRI. 1998. Ecological and Fisheries Status of Reservoirs in India-A recent survey. In: Das, Manas Kr. (ed.). *The Inland Fisheries News*, Newsletter of CIFRI. 3 (1) pp 1-2.

Costa- Percy, B. A. 1998. Constraints to the sustainability of cage aquaculture for resettlement from hydropower dams in Asia: An Indonesian case study. *Jr. Env. and Dev.* Paper available on internet.

Costa-Percy, B. A., and Soemarwoto, O. 1987. Proliferation of Asian reservoirs: The need for integrated management. Naga, The ICLARM quarterly, 10, 9-10.

Devi, Sarla, B. 1997. Present status, potentialities, management and economics of fisheries of two minor reservoirs of Hyderabad Ph. D. thesis, Osmania University, Hyderabad.

De Silva, S. S. 1988. The reservoir fishery of Asia. In S. S. De Silva (ed.) *Reservoir fishery management and development in Asia*. Ottawa, Canada: International Development Research Centre. Pp 19-28.

De Silva, S. S. 1992. Reservoir fisheries of Asia. Ottawa, Canada: International Development Research Centre.

Dixtulu, J. V. H. 1999. Render Indian Reservoirs Sustainably Fishful. *Fishing Chimes.* 19(2): 5 -7.

Dwivedi, S. N. 1980. Reservoir fisheries for Rural Development: New Polices and Technologies. *India Today & Tomorrow.* 8(4): 156-159.

Gorge, Ninan, 2001. Fishery management practices of the major reservoirs in Madhya Pradesh. Paper presented during Nat.

Sem. On Riverine and Reservoir fisheries- challenges and strategies, May 23 to 24, 2001, Cochin.

Jaya Raju, P. B. 2000. Ecological wonders of India: (Ecological degradation due to intensive fish culture practices in lake Kolleru, Andhra Pradesh, (India). *Hydrosphere*, No. 6. IAAB, Hyderabad. pp 40-46.

Jhingran Arun, G. and V. V. Sugunan. 1990. General guidelines and planning criteria for small reservoir fisheries management. p. 18. In: Jhingran Arun G. and V. K. Unnithan (eds). *Reservoir Fisheries in India*. Proc. of the Nat. workshop on Reservoir Fisheries, 3-4 January 1990. Special Publication 3, Asian Fisheries Society, Indian Branch, Mangalore, India

Jhingran, V. G. 1985. Fish and Fisheries of India. Hindustan Publishing Corporation (India), New Delhi: 106, 171, 191.

Mc Cully, P. 1996. Silenced rivers. The ecology and policies for large dams. London: Zed Books.

Mohan Braj. 2001. Economic Status of Fishermen of Malampuzha Reservoir. Paper presented during the Nat. Sem. on Riverine and Reservoir Fisheries-Challenges and Strategies, May 23 to 24, 2001, Cochin.

Pathak, S. C. 1990. Harnessing reservoirs for increasing fish production. p 9-12. In: Jhingran Arun G. and V. K. Unnithan (eds). *Reservoir Fisheries in India*. Proc. of Nat. Workshop on Reservoir Fisheries, 3-4 January 1990. Spl. Publ. 3. Asian Fisheries Society, Indian Branch, Mangalore, India.

Piska, Ravishanka. 1999. Fisheries and Aquaculture. Lahari Publications, Hyderabad.

Piska, Ravishankar and Divakara Chary, K. 2000. Impact of trophic nature of reservoir on the reproductive capacity of catfish Mystus bleekeri (Day). *Ecol. Env. & Cons.* 6(4): 447-452.

Postel, S., Daily, G., and Ehrlich, P. 1996. Human appropriation of renewable freshwater. *Science* 271: 785-788.

Raha, K. and Sarkar, M. K. 1980. Salient Features of the Gomti Reservoir in Tripura. *India Today & Tomorrow* 8 (4): 178-180.

Raghavachari, M. 1983. Reservoir Fishes in India: Some Issues and Problems. In: U. K. Srivastava & M. Dharma Reddy (eds).

Fisheries management in India: Some Aspects of Policy Management. Concept Publishing Company, New Delhi, pp 459-461.

Raghavachari, M. and Rao, S. S. 1984. Development of reservoir fisheries in India: some issues and recommendations. 213-250. In: Srivastava, K. and S. Vathsala (eds). *Strategy For Development of Inland Fishery Resources In India: Key Issues in Production and Marketing.* Proc. of Nat. Workshop on Dev. of Inland Fish Resources held on November 1-3, 1983 at Ahmedabad.

Ringangaonkar, A. M. and M. G. Bhrushundi. 1980. Deforestation of Reservoir Basin: a new indigenous manual device. *India Today & Tomorrow* 8 (4): 172.

Sakhare, V. B. 1999. Fisheries of Yeldari Reservoir, Maharashtra. *Fishing chimes,* 19 (8): 45-47.

Sakhare, V. B. 2001. Ichthyofauna of Jawalgaon reservoir in Solapur district of Maharashtra. *J. Aqua. Biol.* 16 (1&2): 31-33.

Sakhare, V. B. 2001. Reservoir Fisheries in Solapur District of Maharashtra. *Fishing Chimes,* 21 (5): 29-30.

Sakhare, V. B. 2003. Studies on some aspects of fisheries management of Yeldari reservoir, Maharashtra. Ph. D. thesis submitted to Swami Ramanand Teerth Marathwada University, Nanded (Maharashtra).

Sakhare, V. B. 2005. Water quality of Hingni (Pangaon) reservoir and its significance to fisheries. *Advances in Limnology* (ed. S. R. Mishra, Dhar), Daya Publishing House, New Delhi, pp 231-235.

Sakhare, V. B. 2005. Ecology and fisheries of Manjara reservoir in Maharashtra. *Fishing Chimes.* 25 (3): 42-45.

Sakhare, V. B. and Jetithor, S. G. 2011. Fisheries of Harni (Katgaon) Reservoir, Maharashtra. *Fishing Chimes.* 30 (10&11): 98-101.

Sakhare, V. B. and Joshi, P. K. 2002. Ecology and ichthyofauna of Bori reservoir in Maharashtra. *Fishing Chimes.* 22 (4): 40-41.

Sakhare, V. B. and Joshi, P. K. 2002. Ecology of Palas-Nilegaon reservoir in Osmanabad district, Maharashtra. *J. Aqua. Biol.* Vol. 18 (2): 17-22.

Sakhare, V. B. and Joshi, P. K. 2003. Reservoir fishery potential of Parbhani district of Maharashtra. *Fishing Chimes.* 23 (5): 13-16.

Sakhare, V. B. and Joshi, P. K. 2004. Present status of reservoir fisheries in Maharashtra. *Fishing Chimes.* 24 (8): 56-60.

Selvaraj C, and Murugeson, K. 1990. Management techniques adopted for achieving a record fish yield from Aliyar reservoir, Tamil Nadu. P 86-96. In: Jhingran, Arun G. and V. K. Unnithan (eds). *Reservoir Fisheries in India.* Proc. of the Nat. workshop on Reservoir Fisheries, 3-4 January, 1990. Spl. Pub. 3, Asian Fisheries Society, Indian Branch, Mangalore, India.

Shetty, H. P. C. 1990. Forward. *Proc. of Nat. workshop on Reservoir Fisheries,* 3-4 January, 1990. Spl. Publ. Asian Fisheries Society, Indian Branch, Mangalore, India

Singh, Gurucharan. 2001. Status of Development of fisheries of Pong Reservoir (Himachal Pradesh). *Fishing chimes,* 21 (1): 88-90.

Sinha, M. 1998. Policy Options for Integrated Development of Reservoir Fisheries: From Production to Marketing. *Fishing chimes.* 18 (1): 54-59.

Sreenivasan, A. 1989. Principles of ecology and fisheries management of manmade lakes. In: *Conservation and Management of Inland Capture fisheries resources of India.* (eds): Arun G. Jhingran & V. V. Sugunan, Inland Fishery Society of India, Barrackpore. pp96-105.

Sreenivasan, A. 1998. Intrgrated Development of Reservoir Fisheries of India: Production to Marketing. *Fishing chimes.* 18 (1): 60 –63.

Srivastava, U. K., D. K. Desai, V. K. Gupta, S. S. Rao, G. S. Gupta, M. Raghavachari and S. Vatsala. 1985. *Inland fish marketing in India-Reservoir Fisheries,* Vol. 4 (A&B). Concept Publishing Co; New Delhi, pp (A) 403 & (B) 1184.

Sugunan, V. V. and Yadava, Y. S. 1991 b. Feasibility studies for fisheries development of Nongamahir reservoir. CIFRI, Barrackpore, pp 30.

Sugunan, V. V. 1995. Reservoir Fisheries of India. *FAO Fisheries Tech. Report 345.* Daya Publishing House, Delhi.

Sugunan, V. V. 1990. Reservoir fisheries management. In: Sugunan V. V. and U. Bhowmick (eds) *Technologies for inland fisheries development. Central Inland Capture Fisheries Research Institute, Barrackpore, India,* pp 153-164.

Sukumaran, P. K. and Das, A. K. 2001. Status of plankton in some freshwater reservoirs of Karnataka. Paper presented during the *Nat. Sem. on Riverine and Reservoir Fisheries-Challenges and Strategies*, May 23 to 24, 2001Cochin

Valsangkar, S. V. 1980. Economic rehabilitation of fishermen in Yeldari reservoir. *India Today and Tomorrow*, 8(4): 162-163.

Valsangkar, S. V. 1993. Mahseer fisheries of Koyana river (Shivaji Sagar) in Maharashtra: scrap to Bonanza. *Fishing chimes*, 12 (10): 15-19.

Chapter 3

Estuarine Fisheries

An estuary may be defined as a region where river water mixes with seawater and considerably dilutes it. It is the most variable and dynamic aquatic ecosystem, which is influenced by factors like water movements, tides, mixing of sea and river water, salt load and salinity regime. India has an estimated estuarine waters spread area of 2.6 million hectare with the support of mighty estuaries on east and west coasts.

Diversities in salinity regimes and other physico-chemical conditions due to varying intensities to freshwater discharges in different seasons and tidal fluctuations throughout the year bring about biological alterations in the estuarine ecosysem. The faunistic production of adaptable characteristics can withstand the dynamic stress in estuarine waters and thus only a few species of high commercial value are dominating the catches from open waters and also grown in impounded environments.

Open Estuarine System

Hooghly–Matlah Estuarine System

The Hooghly-Matlah estuarine system located in West Bengal between latitude 21°-23°N and longitude 88°-90°E. The Hooghly-Matlah estuarine complex has the largest deltaic region in the world. The total approximate area of this estuarine system with its innumerable tributaries and network of creeks is 2340 sq. km. The mangrove vegetation grown in the deltaic region harbours a potential grazing ground for the estuarine fishes and prawn species. The estuarine sources are highly productive. About 50,000 tonnes of fish are landed annually, 70 per cent being from high salinity zone. Maximum sustainable yield has been estimated to be 3, 56,741 t that is about 200 kg less than what was landed in 1996-97.

Different types of fishing gears are operated in Hooghly-Matlah estuary. Bulk of the total fish landings are contributed by the bag net and the percentage contribution by this type of gear ranges between 70 and 80.

Of special importance in the Hooghly-Matlah estuarine system is the so-called migratory fishing, practiced during winter in the lower zone. Because of the prevailing calm weather in the coastal belt, the fishermen engaged in fishing in upper zone of the estuary migrate to the lower zone for intensive fishing. In the winter, fishing centers like Frazerganj, Jamboodwip, Kalistan etc. the average catch per unit of effort ranges between 29 and 156 kg (1987-88) during November to January, as against 2.6 kg obtained in the upper and lower reaches during the whole year. The important species in bag net catches during winter migratory fishing are *Horpodon nehereus, Settipinna* spp, *Trichiurs* spp and prawns mainly *Parapenaeopsis* spp.

In early seventies, after the construction of Farakka Barrage an improved level of fishery in the Hooghly estuary was indicated with the release of freshwater into the system through the Barrage and the yield in recent years has been on the higher side *i.e.*, around 2000 t per year. The yields of purely marine forms (*P. indicus, T. jella, C. ramcarati* and *L. calarifer*) have markedly deceased in the lower zone while fisheries of some species like *P. pama* and *P. paradiseus* which were in declining trend have shown a sign of improvement in lower and gradient zone of the estuary.

Wide fluctuation in hilsa catch is an interesting characteristic of Hooghly estuary. However, the landing of the species which remained 1457 tonnes/year during 1966-67 to 1974-75 showed a sharp rise of 2126.2 tonnes/year after the farakka discharge (1975-78) and has come down again to 1781.3 tonnes/year in recent years (1984-87). The recent upward trend in hilsa landing can be attributed to the increased effort and mechanization of the boats.

Mahanadi Estuarine System

The Mahanadi estuary lies in Cuttack and Puri districts of Orissa and drains into the Bay of Bengal. The tidal influence extends up to about 42 km from the mouth of the river. As an estuarine complex including the brakishwater lagoons of about 12000 ha, Mahanadi estuarine system covers about 30,000 ha area with an estimated annual production of about 550 tonnes/year.

The fishery of this estuary includes clupeoids, mullets, and prawns. The important fishes present are *Hilsa ilisha, Nematolosa nasus, Sardinella spp., Mugil cephalus, M. cunnesius, M. parsia, Lates calcarifer, Sillago spp., Epinephelus spp., Malabaricus spp., Ambassis spp., Leiognathus spp., Lutianus johnii, Datnoides quadrifasciatus* and *Sparus berda*. The 1959-60 survey reported 6900 fishermen actively engaged in fishing activity using 1700 boats and 56700 nets in this estuarine system.

Estuaries of the Peninsular India

Godavari Estuarine System

It is the major estuary in peninsular India. It has an area of about 18000 hectares. Goutami is the main source of the estuarine complex in which tidal influence extends only up to 40-50 km upstream from the mouth region. Formation of sand bars in the estuarine mouth restricts the entrance of tidal waters. Detailed studies on zooplankton by Ganapati and his colleagues during 1958-64. About 12 species belonging to 7 genera of mysids, 60 species of copepods apart from 17 genera of medusa and other groups such as isopods, amphipods, siphonophores and ctenophores were encountered. The ecology of the bottom invertebrate fauna with special emphasis on polychaetes, molluscs and echinoderms were studied by several workers. A total of 84 species of polychaetes excluding serpulids, 115 species of gastropods, 57 species of pelecypods and 9 species of echinoids have been identified. The total production from Godavari estuary is estimated to be about 5000 tonnes. Mullets and prawns form the major catch of the system. *M. monoceros, M. dobsoni, M. affinis, M. brevicornis, P. monodon,* and *P. indicus* are the important prawn species available in lower reaches of the estuary.

About 200 species of fishes have been recorded from the estuary, of which 72-80 per cent are nearly euryhaline, 12-20 per cent almost marine and 15 per cent freshwater in origin. Mullets form nearly 2/3rd of the total fish landing and are represented by *M. cephalus, M. speigleri, M. dussumieri, L. troschelii, L. oligolepis,* and *L. seheli.* The other fishes of commercial importance are *Pristiopama hasta, Leiognatus sp, Gerres filamentous, Caranx sp, Sillago sihama, Lates calcarifer* etc.

Other Estuarine Systems

Other than Godavari estuarine system there are eight small estuaries in peninsular India namely Vasishtha, Vainatheyam, Adyar, Karuveli, Ponniyar, Godilam, Paravan, Vellar, Killai and Coferoom.

The 40 km long the Adyar river meets the Bay of Bengal at Adyar within the southern limits if Madras city. The tidal influence is felt in the river up to a distance of 6.4 km from its mouth. The estuary is connected to the sea only during monsoon and post monsoon months. When freshwater discharge reduces by the end of January, sand bar is formed at the river mouth and the estuary gets totally cut off from the sea. Common phytoplanktons of the Adyar estuary are represented by *Microcystis, Scenedesmus, Oscillatoria, Merismopedia, Asterionella, Coscinodiscus, Campylodiscus* and *Skeletonema.* Zooplanktons are not abundant, except *Brachionus rubens* through out the year. Among important fishes the estuary is having mullets, *Sillago sihama, Tilapia mossambica,* catfishes and prawns.

The Vellar river drains into the Bay of Bengal at Port Novo. The total length of the river is 480km. The Vellar estuary is located in between latitude 11°-29'N and longitude 79°-46'E. The estuary covers an area of 262 ha. The important fishes of the estuary are *Mugil* spp, *Leiognathus* spp, *Sardinella gibbosa, Johnius* spp, *Polynemus tetradactylus, Engraulis malabaricus,* catfishes and prawns and shrimps. According to Thomas and Raffi (2001) the temperature of estuarine water ranged between 30.5 and 32°C, pH between 8.2 and 8.7 and salinity 13.2 and 82.5 ppm. Maximum value of organic carbon (5.246 mg of C) was recorded in the post monsoon season and the minimum (2.923 mg of C) with a range of 0.69 to 4.83 mg of C during the premonsoon season. The premonsoon recorded the peak values of Cu (110.6 ppm), Fe (48482.1 ppm) and Mn (941.3 ppm); summer recorded increased levels of Cd (129.6 ppm) and Zn (187 ppm). Cr, Ni, Pb and Hg were found in higher levels during monsoon with the values of 460.7 ppm, 579.3 ppm, 194.8 ppm and 535.7 ppm, respectively. The minimum values of Cd (290.0 ppm), Ni (327.6 ppm), Zn (99.7 ppm), Fe (42692.3 ppm) and Mn (871.5 ppm) were recorded in post-monsoon. Monsoon recorded 11.27 ppm of Cu and premonsoon 40.26 ppm, 10.7 ppm and 5.6 ppm of Cr, Pb and Hg as minimum values, respectively.

Estuaries of the North West Coast

Narmada, Tapti and Mahi are three main estuaries in west coast.

Narmada Estuarine System

River Narmada is having an estuarine stretch of about 120km which ends in Gulf of Cambey. Gradual decline in tidal ingress is being observed due to development of sand bars at the mouth region of the estuary. However, from ten years records during 1973-82 the average annual production of the estuarine system has been estimated to be about 4000 tonnes in which as a group hilsa contributed (40-45 per cent average 1662.4 tonnes) followed by mullets (15-20 per cent average 686.95tonnes) and prawns (5-8 per cent average; 286.55 tonnes). Miscellaneous species contributed 30-38 per cent (average 1520 tonnes) of the total landings. The fishing potential estimated in a survey during 1986-87 indicated that 2884 fishermen are actively engaged in fishing activities in the system with 103 boats (non mechanised) and about 28000 nets. Gill nets are in maximum use (26414 nos.) in the lower zone of the estuary. Hilsa forms the major catch of these gill nets (80-85 per cent). Cast nets are mainly used for prawn fishing. Besides these, the seedling of freshwater giant prawn (*M. rosenbergii*) form an important fishery resource in mixoligohaline and limnetic zones of the system.

On account of a cascade of dams under construction on the Narmada river system with the world Bank aid the entire ecosystem and its productivity are bound to be greatly affected. Pre-impoundment surveys have been initiated by the Central Inland Fisheries Research Institute (CIFRI), Barrackpore (West Bengal) in 150 km stretch of the river Narmada in the state of Gujarat.

Chapter 4
Fisheries of Lakes

Fisheries of Pulicat Lake

Pulicat lake is second largest lagoon on east coast spreading in the Nellore district of Andhra Pradesh and the Chingleput district of Tamil Nadu where it is connected with Bay of Bengal by narrow mouth near the pulicat village. In length it is 60 km and the width varies from 1-18km with total area of 77700 ha but with water spread area of 36900 ha rest being dry and becoming marshy during monsoon. Geographically, the lake lies between the latitudes 13°26′ to 13°43′N, and longitudes 80°03° to 80°18′E. Topographically, the lake has been steadily shrinking both in its water spread area as well as in its depth, due to monsoon siltation and summer evaporation. What was about 461 km² during the late Dutch period, (prior to the 19th century A. D.,) and 350 km² during the 20th century, is now according to GIS techniques, just about 267 km² (26,848 ha) since 2004, having shrunk from 289 km² (28,759 ha) of 1967, a shrinkage of about 22 km2, in 37 years (Jayanthi *et al.*, 2007). Employing radio carbon (^{14}C) technique, Caratini (1994) estimated that this lake has been silting up at the rate of about one meter per century, so that projecting it forward, this lake may get totally silted up, within the next half a century, if it is not restored. This increased loss of depth has regarded the lake unfit for permanent inhabitation by several marine species of fish. This siltation has closed up the course and the mouth of the rivulet Aarni in the south, as well as blocking the Buckingham canal at several spots, obstructing the free flow of water into the lake and thus denying the recruitment of biodiversity and fish seed from the several estuaries, back waters and the sea with which the Buckingham canal is inter-connected. Anthropogeneically, developmental activities like the several industries, around Ennore in the south (North Chennai), add chemical, oil and thermal pollution, along with fly-ashy slurry from the two thermal power plants, and the dredged out silt from the Ennore megaport, all observed to be the most major threats for the survival of fisheries and water birds in the southern region of the lake, in recent times. It is connected to sea by a narrow mouth, whose position, depth and width vary with season. Two seasonal rivulets-Rayala vagu and Kalangi river drain into the lake. In the lake, there are several islands the important being Pernadu, Venadu and Irakkam. The average amplitude of the tide at lake mouth is only 25 cm and its influence is felt up to 6 to 10 km from the mouth. The average depth is 1.5 m. During the period 1970-77 total catch varied from 759.6 t (1975) to 1371.4 t (1972) yield ranging from 20.6 to 37.2 kg/ha; of the water spread area.

About 400 species of macrofauna have been described from the Pulicat lake (Sanjeeva Raj, 2006). Phytoplankton is richer than zooplankton both in diversity as well as in populations, but both are unfortunately declining today, due to various ecological threats (Sanjeeva Raj, 2011). The lake classified as 'mesotrophic' type (Raman *et al.*, 1975).

The lake is surrounded by nearly 66 fishing villages,52 of which are on the lake-side, 22 in Andhra Pradesh and 30 in Tamil Nadu. The rest of the villages are on the sea-side in Tamil Nadu (Sanjeeva Raj, 2010a). Prior to the tsunami-2004, there were a total of 43,022 fisherfolk of whom 10730 were active fishermen.

There is no organized data collection of the capture fisheries on the Pulicat lake. During 1970 prawns accounted for 566.5 t of which the main contributors were *Penaeus indicus* (335.9 t), *M. monoceros* (76.1t) and *M. dobsoni* (54.4t). Mullets accounted for 214.6t the main species being *M. cephalus* (136.2t). Perches amounted to 87.4t. Clupeids accounted for 125.6t, the main species being *Nematolosa nasus* (70.2t), crabs were represented by 4 *Neptonus pelagicus* (70.2t) and *Scylla serrata* (48.8t).

Among the gears used Suthuvati (stake net) accounted for about 25 per cent of total catch, Badivalai (shore seine) 13 per cent, Kondavalai (dragnet) 12 per cent and Kalluvalai (stake net) 11 per cent.

Mullets and prawns are the major fisheries of the lake. Fishes like *Mugil cephalus, M. macrolepis, Sillago sihama, Gerres oyena, Nematalosa nasus* and *Chanos chanos*; prawns like *Penaeus semisulcatus, P. monodon, P. indicus, Metapenaeus monoceros* and *M. dobsoni*; crabs like *Scylla serrata* and *Portunus pelagicus* and oyster like *Crassostrea madrasensis* are also found in the lake.

Fisheries of Chilka Lake

The Chilka lake located along the east coast of India between the latitudes 19028'-19024'N and longitudes 85067' and 85035' E is a pear shaped brakishwater lagoon in between the district of Puri, Kaurda and Ganjam has an average water spread area of 1050 sq. kms. It covers an area of 1165 sq. kms. in monsoons and 906 sq. kms. in summer.

The lake is fed with main two tributaries of river Mahanadi called river Daya and river Bhargavi at the northern side and connects with the Bay of Bengal near Arakhakuda village through a zig zag long outer channel of 35 kms long. Similarly the lake has connected with the Bay of Bengal at the southern end near Purunabandha village through 16 kms long canal called as 'palur canal'. As a result of this, the entire lake becomes a freshwater during monsoons. This peculiar phenomenon change provides a congenial features for survival of freshwater flora and fauna up to December and afterwards the salinity gradually increases touching maximum in July.

The depth of the lake is quite uneven, the northern part being shallowest (47-134cm) and the southern and some parts of the central sectors are comparatively deep (102-300 cm). The water level remains minimum during April to June reaching highest during the monsoon period. The transparency of water showed comparatively high values (70-140cm) all along the western bank low values (5-25 cm) were observed on the eastern bank of the northern and central sectors during monsoon period. Annual rainfall at Chilka is recorded at 114-117 cm. The water temperature of the lake varies from 17.5°C in January to32°C in July. The salinity shows sharp annual cyclic changes in the range of 0.13 to 36.02 per cent. The salinity starts rising from November and reaches high values in April-June period in both the central and northern sectors. In the southern sector, however, the fluctuation of salinity is comparatively less pronounced. Dissolved oxygen ranged between 2.6 to 15.6 ppm and total alkalinity from 25.8 to157 ppm. pH was in the range of 6.8 to 9.7. Higher values of phosphates, nitrates, silica and iron occurred in monsoon months while that of total alkalinity in summer months.

The Chilka lake is heavily infested by the aquatic weeds such as *Phanerogams potamogeton pectinatus, Najas fareolata, Halophila ovata, Ruppia martitima* and *Halopila baccavii*. The algae in the lake include the marine species *Gracilaria edulis* and *G. verrucosa* and the fresh species *Hydrilla* and *Vallisneria spiralis* in the southern and central sectors are of economic importance.

The faunal diversity comprises few species of one sea anemones, thirty-seven species of nematodes, one species of polychaetes, seven species of stomotopods, twenty-eight species of crabs, sixty species of mollusca, sixty-nine species of fishes, 150 species of birds, and one species of dolphin.

The lagoon had been facing problems like siltation, shrinkage of area, chocking of the inlet channel as well as shifting of the mouth connecting the sea, decrease in salinity, proliferation of invasive freshwater species, decrease in fish productivity and an overall loss of biodiversity. Due to changes in the ecological characters of the lagoon it was placed in the Motreux record in the year 1993. While comparing the data contained in the pioneering studies on Chilka done during the period 1915 to 1924, with the data generated by way of extensive survey carried out by Zoological Survey of India during 1985-87, it was opined that lagoon ecosystem was tending towards freshwater ecosystem and warranted urgent restoration measures.

Considering the sensitive nature of the lagoon ecosystem, extensive hydrological studies were carried out by commissioning the services of experts from reputed institutes of the country to identify the best options to restore its ecosystem. Based on the findings of the extensive studies, several ameliorating measures were recommended by these institutes and the Chilka Development Authority (CDA) take steps to execute them to restore its ecosystem.

Table 4.1: Fish Species in Chilka Lake

Sl.No.	Species	Family	Habitat
1.	*Ailia coila*	Schilbeidae	pelagic
2.	*Ambassis commersonii*	Ambassidae	benthopelagic
3.	*Anabus testudineus*	Anabantidae	demersal
4.	*Aplocheilus panchax*	Aplocheilidae	benthopelagic
5.	*Bagarius bagarius*	Sisoridae	benthopelagic
6.	*Caranx sexfasciatus*	Carangidae	reef associated
7.	*Catla catla*	Cyprinidae	benthopelagic
8.	*Chanda nama*	Ambassidae	bethopelagic
9.	*Channa punctata*	Channidae	benthopelagic
10.	*Channa striata*	Channidae	benthopelagic
11.	*Chanos chanos*	Chanidae	benthopelagic
12.	*Chela laubuca*	Cyprinidae	pelagic
13.	*Clarias batrachus*	Clariidae	demersal
14.	*Crossocheilus latius*	Cyprinidae	benthopelagic
15.	*Cynoglossus lingua*	Cynoglossidae	demersal
16.	*Echeneis naucrates*	Echeneidae	reef associated
17.	*Elops machnata*	Elopidae	pelagic
18.	*Esomus dancricus*	Cyprinidae	benthopelagic
19.	*Etroplus suratensis*	Cichlidae	benthopelagic
20.	*Gazza minuta*	Leiognathidae	demersal
21.	*Gerres filamentosus*	Gerreidae	demersal
22.	*Gerres setifer*	Gerreidae	demersal
23.	*Glossogobius giuris*	Gobiidae	demersal
24.	*Gudusia chapra*	clupeidae	pelagic
25.	*Heteropnestus fossilis*	Heteropneustidae	demersal
26.	*Hilsa kelee*	Clupeidae	pelagic
27.	*Labeo ariza*	Cyprinidae	bentopelagic
28.	*Labeo calbasu*	Cyprinidae	demersal
29.	*Labeo rohita*	Cyprinidae	benthopelagic
30.	*Lates calcarifer*	Centropomidae	demersal
31.	*Liza macolepis*	Mugilidae	demersal
32.	*Liza parsia*	Mugilidae	demersal

Contd...

Table 4.1–*Contd...*

Sl.No.	Species	Family	Habitat
33.	*Liza tade*	Mugilidae	demersal
34.	*Macrognathus aral*	Mastacembelidae	benthopelagic
35.	*Mastacembelus armatus*	Mastacembelidae	demersal
36.	*Megalops cyprinoides*	Megalopidae	pelagic
37.	*Mugil cephalus*	Mugilidae	benthopelagic
38.	*Mystus cavasius*	Bagridae	demersal
39.	*Mystus gulio*	Bagridae	demersal
40.	*Mystus vittatus*	Bagridae	demersal
41.	*Nematalosa nasus*	Clupeidae	pelagic
42.	*Notopterus notopterus*	Notopteridae	demersal
43.	*Ompak bimaculatus*	Siluridae	demersal
44.	*Pangasius pangasuis*	Pangasiidae	benthopelagic
45.	*Plotosus canius*	Plotosidae	demersal
46.	*Plotosus lineatus*	Plotosidae	Reef associated
47.	*Pristis pectinata*	Pristidae	demersal
48.	*Pterois radiata*	Scorpaenidae	reef associated
49.	*Puntius sophore*	Cyprinidae	benthopelagic
50.	*Puntius ticto*	Cyprinidae	benthopelagic
51.	*Puntius vittatus*	Cyprinidae	benthopelagic
52.	*Rasbora daniconius*	Cyprinidae	benthopelagic
53.	*Rasbora rasbora*	Cyprinidae	benthopelagic
54.	*Rhinomugil corsula*	Mugilidae	pelagic
55.	*Salmostoma bacaila*	Cyprinidae	benthopelagic
56.	*Sardinella melannura*	Clupeidae	pelagic
57.	*Scoliodon laticaudus*	Carcharhinidae	demersal
58.	*Silonia silondia*	Schilbeidae	demersal
59.	*Sphyraena putnamae*	Sphyraenidae	reef associated
60.	*Tenualosa ilisha*	Clupeidae	pelagic
61.	*Terapon jarbua*	Tetrapontidae	demersal
62.	*Tetradon cutcutia*	Tetradontidae	demersal
63.	*Wallago attu*	Siluridae	demersal
64.	*Xenentodon cancila*	Belonidae	pelagic

A two day international workshop on 'Eco-restoration of Chilka lagoon' was organized by CDA at Bhubaneswar. The workshop was held from 19 to20 January, 2002. The prime objective of the workshop was to conserve and restore the lagoon's ecosystem and at the same time facilitate its sustainable use for promoting economic well being of the local community dependent upon the lagoon for its livelihood and to generate clear recommendations regarding actionable strategies in all needed areas for conservation. Some important recommendations made in the workshop were:

1. In the context of the increase in tidal flow and the salinity observed in the main lake, the variation has to be monitored for the coming years and the real impact can be assessed there upon.

2. The implementation of efficient river management system may be initiated.

3. The study on the littoral drift in ocean side and by passing around the new mouth can be undertaken.

4. Remote sensing application can be effectively used to enhance the strength of the field data collection.

5. Detailed mapping and survey of biodiversity and continuous monitoring may be undertaken.

6. Measures to protect catchment and to prevent sedimentations may be taken.

7. Regular monitoring of floral and faunal diversity necessary may be undertaken.

8. All Dolphin carcasses be preserved and studied.

9. Khanda fishing in Magarmukh and Nalaban be stopped including capture of juveniles. Palur canal be deepened up to the melting point with the open sea.

10. Prawn culture and conversion to agriculture be stopped.

11. Sustainable fish yield and balancing between ecological and economic aspects should be aimed at.

12. Gender issues to be addressed for sustainable fisheries development.

Fisheries of Kolleru Lake

Kolleru lake is the second largest freshwater ecosystem in peninsular India. It is located between 16°32' latitudes of 16°47'N and longitudes of 81°2′ and 81°4′E in the state of Andhra Pradesh. The lake was formed as a natural depression between river Godavari on the north and river Krishna on the south near their tail-end region, before they open into Bay of Bengal. Several irrigation canals originating from the dams on river Godavari at Dhowlaiswaram and river Krishna also open into the lake. Lake also receives water through a number of large and medium sized, seasonal rivulets and streams that drain the surrounding upland areas. The lake is connected to Bay of Bengal through a narrow natural channel Upputeru, about 62 km long.

The water spread area of 901.3 sq. km. at maximum and 135 sq. km at 1m (M. S. L.). The total catchment area of the lake was estimated to be about 4700 sq. km. Depending upon the seasonal inflow the depth of lake varies between 1m to 4m. During summer the marginal shallow areas of the lake become completely dry.

The lake area receives annual average rainfall of 99.95 cm. The relative humidity varies between 64 per cent in June and 83 per cent in February. The lake experiences brackishwater conditions in the months when the inflow of freshwater into the lake is low. The saline water reaches up to the middle regions of the lake, while in the northern parts freshwater conditions extist. On the onset of southwest monsoon in the mid seasonal drains and streams.

Kolleru lake supports very rich aquatic flora and fauna. The lake is a eutrophic body with huge quantities of submerged, floating

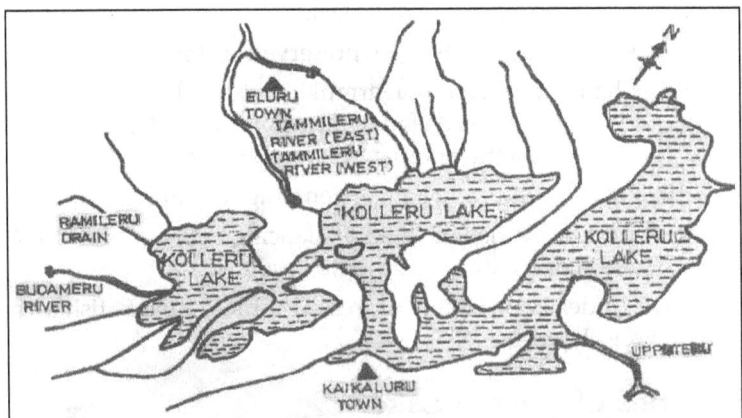

Figure 4.1: Map of Kolleru Lake, Andhra Pardesh

and emergent aquatic vegetation. The lakebed is completely covered with rooted macrophytes. The common species are *Ipomoea aquatica, Salvinia auriculata, Eichhornia crussipes* and *Pistia stratiotes.* Among zooplanktons rotifers, chonchostracans, cladocerans, copepods, cyclopoides, harpacticoides and ostracods are dominant.11 species of birds have been recorded from lake area.

Near about 105 species of fish and 7 species of commercially important prawns were recorded so far from the lake, out of which 19 species of fish were encountered in brackish waters in the tidal zone. The lake supports a rich fishery of air breathing fishes. Traditionally live fish captured in the lake area exported to markets in the north-eastern India. The Indian climbing perch (*Anabus sp*) dominate the catch. The commercially important catfish include *Clarias batrachus, Heteropneustes fossilis* and *Wallago attu.* The murrels such as *Channa striatus* and *Channa punctatus* supports a good fishery. Carp landings are represented by *Catla catla, Labeo rohita, Cirrhinus mrigala* and *Puntius sarana sarana.* Other important fish species are *Mugil cephalus, Etroplus suratensis, Etroplus maculates, Mastacembelus armatus armatus, M. pancalus, Macrognathus aculeatus* and *Anguilla bengalensis.* The commercially important prawns are *Penaeus monodon, Penaeus indicus, Metapenaeus monoceros, M. dobsonii, Macrobrachium rosenbergii, M. rude,* and *M. macolmsoni.*

The marginal areas of the lake have been converted into fishponds. World Banks extended financial assistance to farmers for the construction and management of fishponds. In the last decade nearly 24000 hectares area have been converted into the fishponds. This has resulted in the degradation of natural habitat in the lake proper over the years affecting the fishery. The area of the lake has diminished and the breeding areas of the many species are affected.

Dutt and Murthy (1976) studied the fisheries of Kolleru lake for 1970-72, when the ponds were not constructed. During this period the total fish landings were estimated to be around 720 tonnes. Of these fish *Anabus oligolepis* and *A. testudineus* dominated the landings forming nearly 28 per cent. Carps contributed to 18 per cent and murrels to about 16 percent. Rao *et al.* (1987) based on a study of the landings to be around 1800 to 2800 tones. The total landings during the period between 1976 and 1987 increased three fold. It is mainly due to an increase in the fishing activity in the lake.

Table 4.2: Major Landings of Fish (in per cent of the total fish landings) per cent in the Catches from the Kolleru Lake

Major Groups	Before Construction of Fish Ponds (Dutt and Murthy 1976)	After Constrction of Fish Ponds (Rao et al., 1987)
Perches	28.37	35.39
Catfish	13.19	32.37
Murrels	16.61	17.01
Carps	18.75	10.73
Mullets	–	1.38
Spiny eel	22.80	0.99
Eels	1.00	0.14

A comparative study of the major fish groups on the fisheries of Kolleru lake between the periods before and after the changes in ecology suggests perceptible changes in the fisheries of Kolleru lake. Catfish landings when compared to earlier period also increased. The landings of the climbing perches and murrels remained stable, while that of carps declined remarkably. This could be attributable to the decrease in the lake area, where the juveniles are distributed and also the increased capture of juveniles of carps for stocking in the near by fish ponds. Normally the juveniles got recruited from the streams and irrigational channels opening into the lake. Increaseed exploitation of juveniles could be one of the reasons for the decline in fishery.

Fisheries of Dal Lake

The Kashmir valley has a number of freshwater lakes notably Dal lake, Wular Lake, Haigam Rakh, Mirgund and Hokarsar. These lakes are unique in their flora and fauna, their beauty and their biological production. The Dal lake is situated within the Srinagar city lying with a latitude of 34.7°N and longitude of 74.52°E. It is a Himalayan urban lake which is mainly used for tourism. Fishery is of secondary importance. The lake is one of the most beautiful lakes of India. The maximum depth of the Dal lake is around 1.37 meter.

Dal lake is also the birds paradise. The birds such as white breasted Kingfishers, little grebe, common pariah and the grey heron are the common birds.

Table 4.3: Physical Features of Dal Lake

1. Surface area	21 km²
2. Volume	0.00983 km³
3. Mean depth	1.4 meters
4. Maximum depth	1.37 meters
5. Length of shoreline	15.5 km
6. Catchment area	316 km²

The phytoplanktons are represented by chlorophycceae, bacillarophyceae and myxophyeae. About 150 phytoplankton genera have been identified. The maximum population of phytoplankton has been recorded in May to June and minimum in December to January. The secondary peak occurs in the month of September.

Copepoda, rotifera and cladocera are the principal components of the zooplankton. Copepoda forms the largest group, next being rotifera and third cladocera, all showing maximum population in spring and autumn and poor population in summer and winter. Ostracoda, having 2 genera, has been found absent in the late spring, late summer and early autumn.

Thirty seven species of fish have been recorded and important species inhabiting the Dal lake are *Schizothorax esocinus, Schizothorax niger, Schizothorax curuifrons, Labeo dero, Crossocheilus sp, Botia birdi, Nemacheilus kashmirensis, Gambussia afffinis, Glyptothorax, Glyptosternum* etc.

Besides the above mentioned fishes, the exotic fishes like scale carp (*Cyprinus carpio communis*) and mirror carp (*Cyprinus carpio specularis*) constitute the major catch of commercial fishery of Dal lake. Common carp introduced in the year 1957, constitutes about 70 per cent of the total catches where as *Schizothorax* contributes for about 10 per cent and other fishes including catfishes form 20 per cent of total landings.

Cast net is the main fishing gear used in the lake. It is of 6 pieces with 4.8 meters in diameter having mesh size 1 to 2 cm, bar to bar with iron or lead sinkers along the peripheral cord weighing about 5 kg per net. One fisherman operates the net from boat while another fisherman rows the boat. Besides this, scoop net and hook and line are most commonly used by the fishermen. Narsoo, a multiple head spear having 5 to 7 heads, is also used. The nets are fabricated locally by the fishermen, mostly of nylon twine, but cotton twine is also used.

The fishing craft consists of open wooden boat of about 20 feet length and 4 feet center width, having capacity of carrying 2 to 3 persons, besides fish catch of about 50 kg. The boats are mostly constructed from devdar wood.

The factors responsible for pollution in the Dal lake are sewage disposal, supply of nutrients from rice fields, catchment areas, urban surrounding and numerous house boats. The mortality of endemic fish *viz.*, *Schizothorax* and *Botia* is suspected to be due to a large extent to the accumulation of residues of pesticides used for the control of pests of fruits, orchards and floating vegetable gardens in and around the Dal lake and it has been stressed that pesticides and insecticides should be carefully and judiciously used.

The cause of deterioration of lake is due to the excessive inflow of nutrients, overgrazing and encroachment of lake area and wastes from surroundings and houseboats. Removal of excess weeds from the lake is essential. It can be done by using grass cutters or by chemicals, which would kill the weeds but not the fish. The weeds can also be controlled by introducing the Grass carp (*Ctenopharyngodon idella*). It is a hardy, fast growing fish. It may not compete with the existing fish fauna of the lake. Grass carp effectively controls *Azolla, Salvinia, Hydrilla, Najas, Ceratophyllum, Nymphoides, Nelumbo, Vallisneria, Utricularia, Myriophyllum, Potamogeton* etc which form the common weeds in the Dal lake. These weeds not only use up the plant nutrients in the water but also shade up sunlight, causing less production of phytoplankton. This in turn causes less zooplankton production, as the zooplankters are consumers of phytoplankton. Consequently this leads to starvation or low diet of the fish larvae, fry and fingerlings, 50 to 80 per cent of these endemic fish juveniles dying out in the lake. Then again, pollution kills off large numbers of these juveniles. Pollution in Dal lake can be prevented by special arrangements to pipe off the sewage and excreta, which falls into the lake through the flush systems of over 2000 houseboats all over the lake. This sewage may be collected and converted easily into fertilizers.

Fisheries of Loktak Lake

Loktak lake is the largest freshwater wetland in North-Eastern India. It is also known as the only floating lake in the world due to the floating phumdis on it. Loktak is a shallow and somewhat acidic freshwater lake. It is the largest wetland in the North-Eastern region of India and has been referred as the lifeline of the people of Manipur due to its importance in the socioeconomic and cultural life. The lake is situated 38 kms south of Imphal city of Manipur state. It covers an area of about 286 sq. kms at the elevation of 768.5 m located between longitudes 93° 46' and 93° 55' E and latitudes 24° 25' and 24° 42' N. The depth of water during dry season ranges between 0.5 m to 1.5 m. The total water spread area of about 490 sq. kms. was recorded in 1966. Loktak lake plays an important role in the ecological and economic security of the region. The Lake has been the source of water for generation of hydroelectric power, irrigation and water supply. A large population living around the lake depends upon the lake resources for their sustenance. The lake is included on Montreux record in 1993 as a result of ecological problems such as deforestation in the catchment area, infestation of water hyacinth, and pollution.

A number of streams flow directly into Loktak lake. These includes Nambul, Nambol, Thongjarok, Awang Kharok, Awang Khujairok, Oinan, Keinou and Irulok contribute maximum silt load to the lake. The indirect catchment area covers catchments of 5 important rivers *i.e.*, Imphal, Iril, Thoubal, Sekmai and Khuga and is spread over an area of 7157 sq. kms. It is rich in biodiversity and has been designated as a wetland of international importance under Ramsar convention in 1990.

The biodiversity of lake comprises 233 species of aquatic macrophytes of emergent, sumergent, free floating and rooted floating leaf types. The lake also provides refuge to thousands of birds, which belong to at least 116 species.

The lake is known for its unusual infestation with weeds, both dead and live, and also heavy siltation. This infestation piled up over years into various floating shapes, oblong, circular etc, with a depth of about 5 to 6 feet. These are known as 'phums'. They cover about half of the surface of the lake. Phums vary in size and can be as large as one ha. In a way, they have turned out to be a sort of fish aggregation devices that facilitate fish harvesting. Over years, the fishermen dependent on the lake fisheries improvised a certain alteration in the structure of several of the phums. Leaving a rim all round, they have removed the rest of the phum. The truncated area is used as a feeding place for fish which are offered fresh weeds, rice bran and ground nut oil cake powder as feed to attract their aggregation.

The lake is being stocked with fingerlings of carps since 1966. October to March is the season for phum fishing. Fishermen encircle a phum with a net to collect the fish. It is stated that phums yield between 15-40 t of fish annually, besides catches taken through use of hook and lines etc.

Phums get dispersed sometimes moving over long distances within the lake. When this happens, fishermen improvise fresh phums close to their habitations. Another feature one notices is that huts are built on phums and are mostly used as lodges for visitors or for fishermen's use.

In order to utilize the potential of the lake for beneficial purposes, the state government has set up Loktak Development Authority (LDA). The main purpose of the LDA was to set up a hydroelectric project, but beside this LDA is also charged with the responsibility of development of fisheries of the lake. The authority has taken up the work of cleaning the lake from phums and scattered weeds. To facilitate stocking, the authority has set up two major carp hatcheries.

The fish fauna of lake comprises 64 species. Loktak lake serves as the breeding ground for several species of migratory fishes such as *Labeo dero, L. angra, L. bata, Cirrhinus reba* and *Osteobrama belangeri*. These riverine species migrate from the Chindwin Irrwaddy river system in Burma to the upstream areas of Manipur river and breed in various shallowlakes in the valley. These fishes disappeared from the lake since the construction of Ithai barrage which has blocked their migratory route. The fisheries of lake is also supported by *Hypophthalmichthys molitrix, Ctenopharyngodon idella, Cyprinus carpio, Cirrhinus mrigala, Labeo rohita, L. calbasu, L. gonius, Catla catla, Puntius* spp, *Esomus danricus, Heteropneustes fossilis, Ambassis* sp, *Osteobrama cotio, Channa* spp, *Glossogobius* spp, *Notopterus* spp, *Amblypharyngodon mola, Anabus testudineus, Lepidocephalus* spp, *Clarias batrachus* and *Tilapia sp*.

Lokatak lake serves as the breeding ground for several species of migratory fishes such as *Labeo dero, Labeo angra, Labeo bata, Cirrhinus reba* and *Osteobrama belangiri.*

The fish catch is dominated by exotic carps (33 per cent), followed by Indian major carps (21 per cent), minnows (14 per cent) and rest by other forms. The fish production of the lake is 78 kg/ha/yr (Sugunan, 1998).

The hooks used are mainly of no.12 and 11. In each line there are about 5 to 7 hooks tied to different snoods of length 9 to 11 cm. The lines are of synthetic monofilament with size varying from 0.45 to 0.60 mm diameter. The hooks are baited and hurled to a distance of about 15 m from the shore. Such hooks and lines are set in series numbering about 2000 along the banks at a gap of about 0.5 m in between. Each line is wound around an empty can of 10 cm diameter and height of 14 cm for retrieving the hooks and the cans are weighted with stones parallel to the shore at a height of about 40 to 45 cm from the water surface so that they may not be easily swept. This innovation is to prevent the retrieving cans from being carried away by the fishes. As the lines rest on the supporting rope, when the hooked fish drags the retrieving can, it gets entangled in the supporting rope by gravity. After the bait gets disintegrated, the lines are hauled up. The lines which hook catches are pulled slowly and when the fish drags on the opposite direction the line is released slightly to reduce tension in the line. When the fish reaches the shore, it is scooped out of water using a scoop net having a mouth diameter of 65 cm and depth of 40 to 45 cm made with yarn of diameter 1.8 mm.

Another technique of fishing in Loktak lake is the use of pole and line with bait. In this method lightweight bamboo poles of length 2 to 3 m are used with nylon lines of length 5 to 6 m. A no.18 hook is tied at the end of the line and after a gap of 4 to 5 cm; another hook of same size is tied. In between the two hooks, a snider made of lead is loaded and a float of dried stick is tied at a height of 1 to 1.5 m from the bottom hook. A platform is constructed with wooden planks and the four legs are fixed at the bottom of the canal and pole and line is operated from this platform. The bait is in the form of wheat flour mixed with water forming a dough and baited in small pieces of about 3 mm diameter. The fishes caught are kept in a container which is made of bamboo chips and is kept partly immersed in the water. The catch ranges from 0.5 to 3 kg per day per hook and the intensity of catch varies from season to season having a bimodal peak during March-April and August-September and a lean season during January-February.

In ripping method of fishing No.9 or 11 hooks are used. The hooks are tied by means of rolling hitch knot with snood length of 9 to 11 cm and is tied to the end of the line which is about 4 m long, synthetic monofilament twine, with 0.66 mm diameter of which the other end is tied to a pole of length 2 to 3 m. These are operated when fishes exhibit surfacing during morning hours on some cloudy days, which may be due to low concentration of oxygen. At this time fishermen lash the fish quickly with a bunch of hooks holding of the end of the pole. When the hooks rip the fish, the catch is pulled towards the shore and finally scooped out of water with a scoop net.

Dip net of size ranging from 2×3 m^2 to 1.5×2.5 m^2 are also used at few points. The catch ranges from 1 to 5 kg per net per day with a peak during August to January.

Threats and their Impacts

1) Over-exploitation, indiscriminate methods of fishing, extensive growth of phums are responsible for decrease in fish production. Construction of Ithai barrage has interfered with the fish migration from Chindwin-Irrawady river system of Myanmar and consequently brought changes in the catch. 2) Extensive deforestation and unscientific land use practices in the catchment area are responsible for deposition of silt in the lake. 3) The proliferation of phums and aquatic weeds have led to the reduced water holding capacity, deterioration of water quality, interference in navigation, and overall aesthetic value of the lake. 4) Inflow of organo-chlorine pesticides and chemical fertilizers used in the agricultural practices around the lake, muncipal wastes brought by Manbul river that runs through Imphal, soil nutrients from the denuded catchment area and domestic sewage from settlements in and around the lake are responsible for deterioration of water quality.

Fisheries of Sasthamcotta Lake

It is the largest freshwater lake in Kerala state, extending to about 3.42 sq. km. The lake is situated at 09°02'N and 76°37'.'The ancient Sastha temple of the lake is an important pilgrimage center.

It is spring fed lake and is the source of drinking water for half a million people in the Kollam district of Kerala. Some 27 fish species are present. The lake is famous for catches of fishes like *Etroplus suratensis, Etroplus maculates, Ambassis dayi* and *Horabagarus brachysoma*. The water contains no common salts or other minerals and it supports no water plants; a larva called 'cavaborus' abounds and eliminates bacteria in the water, thus contributing to its exceptional purity.

References

Badapanda, H. S. 2002. Planned eco-restoration of Chilka lagoon. *Fishing Chimes*. 22 (4): 20-23.

Caratini, C. 1994. Pulicat: A four century story. The Hindu, Sunday, October 9, p11.

Dixitulu, J. V. H. 2002. On Loktak lake fishery development (editorial). *Fishing Chimes*. 22 (4): 4-5.

Dutt, S. and V. S. R. Murthy. 1976. On the fish and fisheries of the lake Kolleru, Andhra Pradesh. *Mem. Soc. Zool.*, Guntur 1: 17-21.

Jayadev, W. and W. Vishwanath. 2002. Fishing practices at water inlet area of Loktak hydel project. *Fishing Chimes*. 22 (4): 57-58.

Jayanthi, M., Kavitha, N., Ravichandran, P. and Muralidhar, M. 2007. Assessment of impact of brackishwater aquaculture on Pulicat Lake environment and its nearby resources using remote sensing and GIS techniques. *J. of Aquaculture in the Tropics*. 22 (1-2): 55-69.

Jaya Raju, P. B. 2000. Ecological wonders of India: (Ecological degradation due to intensive fish culture practices in lake Kolleru, Andhra Pradesh, India). *Hydrosphere* 6: 40-46.

Raman, K., Ramakrihna, K. V., Radhakrishnan, S. and Rao, G. R. M. 1975. Studies on the hydrobiology and benthic ecology of lake Pulicat. Bull. Dept. Mar. Sci., Univ. Cochin. VII (4): 855-884.

Rao, K. V., R. S. Panwar, M. Rama Krishnaiah, J. B. Rao, T. S. Rama Raju, K. S. Rao and G. Ch. Rao. 1991. Studies on ecology and fisheries of Kolleru lake ecosystem and development of suitable management measures for obtaining sustained fish production. Annual Report, Central Inland Fisheries Research Institute, Barrackpore: 26-30.

Sanjeeva Raj, P. J. 2006. Macrofauna of Pulicat Lake. NBA Bulletin No. 6, National Biodiversity Authority, Chennai, Tamil Nadu, India, p. 67.

Sanjeeva Raj, P. J. 2010a. Management of Pulicat Lake. Divisional Forest Officer, Wildlife Management Division, Sullurpet, pp 47.

Sanjeeva Raj, P. J. 2011. Sustainable fishery development of Pulicat lake. *Fishing Chimes*. 30 (10&11): 67-71.

Sugunan, V. V. 1998. Fishery resources of the north-eastern region and scope for their development. *Fishing Chimes*. 18: 64-71.

Chapter 5
Mahseer Fishery

Mahseer, king among sport fishes is favourite with the anglers not only in India but from distant countries like India, Pakistan, Bangladesh, Myanmar and Sri Lanka. It gives more sport for its size than any other fish. The species of mahseer which are generally found in rivers of India including its peninsular zone are as follows:

Scientific Name	Common Name
Tor putitora (Ham.)	Golden mahseer
T. tor (Ham.)	Turia or tor mahseer
T. mosal (Ham.)	Copper mahseer
T. progeneius (Mc clld.)	Jungha mahseer
T. khudree (Sykes)	Deccan mahseer
T. mussullah (Sykes)	Humpback mahseer
Tor douronensis (Valenciennes)	Mahseer
Tor sinensis (Wu)	Mahseer
Tor tambroides (Bleeker)	Mahseer
Neolissohilus (McClld)	Chocoloate mahseer
Tor zhobensis (Mirza)	Mahseer

Distribution

Putitor mahseer is distributed all along the foothills of the Himalayas from Assam to Himachal Pradesh. The distribution extends up to Poonch Valley of Jammu and Kashmir (Surendranath, 1982). It is a common hill stream major mahseer species of Garhwal (Nautiyal and Lal, 1982). Sehgal (1972) reported the presence of *Tor putitora* in river Yamuna and its tributaries Giri, Bata and Markanda. Its limit of distribution in the Indo-Gangetic drainage of western Himalayas is up to the Hindukush-Kabul-Kohistan watershed (Jhingran and Sehgal, 1978b), extending to the eastern Himalayas through Duars and Terai (Shaw and Shebbeare, 1937).

Tor mahseer more stoutly built than putitor mahseer is also a typical fish for anglers and attains a size of about 1.5m. It occurs along the foothills of Himalayas range from Hindukush-Kabul-Kohistan watershed through Arunachal Pradesh and in most of the river systems and reservoirs in the hilly terrain of Vindhyas and Satpura in Madhya Pradesh. Day (1878, 1889) recorded its distribution throughout India, and its abundance in mountain streams including those which are rocky also as in Sri Lanka. Hora (1940) and Jayaram (1981) reported on natural distribution of this species of mahseer in the east Punjab, Uttar Pradesh, Western Himalayas, Bihar, Darjeeling district, North Bengal, Assam, Eastern Himalayas and Madhya Pradesh. Tor mahseer were reported in adjacent countries of India such as West Punjab of Pakistan, Bangladesh and also China (Hora, 1940 and Jayaram, 1981). It also occurred in abundance in the river Barack of Assam (Hora, 1993b), Aruada of Madhya Pradesh (Karamchandani et al., 1967, Desai 1970), Govindgarh lake (Pisolkar and Karamchandani, 1981), Pond reservoir (Dubey, 1985), in Garhwal, Uttarkashi and in Dehradun valley of Uttaranchal, in the flowing waters of the hilly terrain from Jagadhri to Huthnikund in Ambala district of Haryana, and in the vast stretches of river Beas from above Mandi in Himachal Pradesh to Mirthal in Punjab. It also occurs in the rivers of Pourma and Thailand (Jhingran and Sehgal, 1978). The mosal mahseer occurs in the rivers in and around Eastern Himalayas, Mahanadi basin and Burma waters. It attains size of 1.2m.

The khudree mahseer (*Tor khudree*) also called deccan mahseer is somewhat like tor mahseer and grows up to 1.5m. This is found in the entire peninsular India, South of river Tapti (Kulkarni, 1970).

This species is also available in Uttar Pradesh and Orissa (Mishra, 1959). Jayaram (1981) mentioned the wide distribution of this species in the peninsular India especially Kerala, Karnataka and Maharashtra. CIFRI (1983) reported availability of this species from Rihand reservoir (Uttar Pradesh) and Bhavanisgar reservoir (Tamil Nadu).

Tor mussullah is mostly found in the river systems of peninsular India. Chacko (1948) recorded the species from Hegenakal, Tamil Nadu. Silas (1953) noted the occurrence of this species from Mahabaleshwar lake and Krishna river at Wai (Maharashtra).

Tor douronensis occurs in Thailand and eastwards in Vietnam and towards South in Indonesia. Its occurrence in Chao Malaysia (North Bornio) is also known (Inger and Chin, 1962).

Neolissochilus hexagonolepis is reported from the rivers of Assam. Outside India it occurs in Bangladesh, Mynmar, China and Nepal. It is most abundant in northeast India (Dasgupta, 1994). It is also found in the Cauvery River in Tamil Nadu.

Food

Tor tor is largely a herbivorous form. *Tor putitora* is a omnivorous and largely herbivorous fish. *Tor khudree* takes small fish in certain months, it is predominantly dependent on material of vegetative origin, insect larvae, and molluscs.

Fisheries

Capture fishery in upland water is poorly developed primarily due to difficult terrain and inaccessibility. Excat fish catch statistics of mahseer fishery is wanting for proper comparison but whatever scattered data and reports are available indicate decline in its fishery.

Fish catch data from Gobindsagar (Himachal Pradesh) for 13 years (1974-87) clearly indicate decline of mahseer fishery though total landings have increased from 174t to 538t but contribution of mahseer has declined from 46.8t (20.5 per cent) to 7.5(2.00 per cent). Similarly in Pong reservoir, catch of mahseer came down from 101.5t (20.4 per cent) to 86.9t (10.9 per cent) when catch increased from 498.8 to 797.4t during the 1982-87. Contribution of mahseer in Narmada river during 1959 to1966 also showed declining trend. Catch came down from 23.6 t to 19.8 t. In Bhimtal lake (Uttar Pradesh) mahseer catches dropped from 415 to 385 kg. Two species (*T. putitora* and *T. tor*) also showed similar decline in their catches in three streams of Assam. Their contribution of 42 per cent in 1972 came down to only 0.05 per cent in the total fish catch during 1980. In Maharashtra catch of *T. khudree* in rivers like Bhima, Krishna and Koyana has become a rarity. In south India the situation is in no way brighter. The river cauvery which was at one time the home of khudree had become practically devoid of mahseer. Recently *T. khudree* fingerlings from Lonavala (Pune) were stocked and now anglers are having thrill of angling mahseer as big as 40-45 kg.

The reasons for the decline of mahseer are-destruction of the fish by use of explosives and catching of brood fish while ascending and descending during spawning migration. Another serious handicap the fish suffers in change of their ecosystem due to coming up of multipurpose reservoirs etc. Thus, the biotope is changed as such their breeding grounds are lost there by the auto-stocking is hampered.

Natural Breeding

Many workers have given different breeding habits of the same species of mahseer from different parts of India; hence the controversy of breeding periodicities of mahseer exists (Table 5.1).

Table 5.1: Breeding Season of Different Species of Mahseer

Species	Breeding Season	Source
Tor putitora	Breeding times a season	Thomas (1897)
Tor putitora	Breeding commences with monsoons (In Kumaon region)	Dunstord (1911)
Tor putitora	Several times in a year	Quasim Quayyum (1962) Bhatnagar (1967)
Tor putitora	August to September (Jammu streams)	Sunder and Joshi (1976)
Tor putitora	May to June August to September (River Beas)	Sehgal (1978)
To tor	July to December (River Narmada)	Karamchandani *et al.* (1967)
Tor mosal	November to December (River Mahanadi)	Ahmed (1953)
Tor khudree	July to September (Lonavala reservoir)	Kulkarni (1971)
Tor khudree	July to September	Cordrinton (1946)

In north Indian rivers the mahseer breeds three times a season. In the plains of North India putitor mahseer is reported to breed several times over a greater part of the year. In the Eastern Doons and elsewhere in the Himalayan rivers, however, putitor, tor and mosal mahseers breed same time during August-September only. In the rivers of peninsular India, fed by monsoon floods, mahseer breeds twice –July –September and January-February. The mosal mahseer breed during monsoon (August-September) in Deolal hills and in cauvery river system during winter (November-January). Mosal mahseer breeds after monsoon in November-January in mahanadi.

Fecundity of fish is comparatively low. Karamchandani *et al.* (1967) calculated fecundity of the mahseer as 30,400 ova of a fish 625mm in length while Desai (1973) reported 42,600 ova in 657mm length female. Kulkarni (1978) recorded 20,000 ova from *T. khudree* of 630mm size.

Incubation period is quite long ranging between 80-120 hours. This period is likely to be longer in cold streams of sub-Himalayan region of mahseer namely *T. putiotra* and *T. tor*. Kulkarni (1971) reported incubation period as 80 hours with water temperature of 22-26°C. Tripathi (1977) recorded 80-96 hours for hatching of *T. putitora* eggs at 21°C and Joshi (1982) reported 58-120 hours at 16-25°C. Kulkarni (1980) found for *T. tor* eggs the average hatching period of 82 hours at 24°C. Hence it can be inferred that incubation period for mahseer varies as per water temperature.

Artificial Propogation

Commendable success has been achieved in recent years in breeding mahseer by artificial means. Kulkarni and Ogale (1978) were probably the first to breed *T. khudree* by hormone injection at Lonavala followed by Pathani and Das (1979) in case of *T. putitora*. Tripathi (1978) and Joshi (1982)also made successful attempts to breed *T. putitora* by stripping the brooders at Bhimtal during August-September.

Air Transport of Eggs

In order to facilitate distribution of mahseer seed to distant places efforts were successfully made to transport mahseer egg by air in moist cotton from Mumbai to Bangalore. In this method, fertilized eggs are allowed to harden and develop for about 24 hours, then placed on moist cotton in two layers in perforated plastic boxes and the latter packed in suitable containers. As the minimum hatching period is 60 hours, sufficient time is available for transport over long distances. Success of such method will enable transportation of fertilized eggs by air services to any place in the country or even abroad. On reaching the destination the eggs can be hatched in the normal manner and the resultant fry and fingerlings released into desired sheets of water (Kulkarni, 1979).

Conservation

Indiscriminate capture, use of destructive methods of fishing like chip netting, dynamiting etc have caused a sharp decline in the fishery of mahseer. It is high time its conservation is made before it reaches the stage of non-return. Some of the conservation methods are –the breeding grounds of mahseer are completely closed for fishing at least for a certain season *i.e.*, closed season during breeding period. Certain stretches of rivers may be declared as sanctuaries and enforcement of size limit and annual catch limits may be judicially made. The quantitative improvement of streams is possible by transplanting of farm-reared stock. Salvaging of mahseer seed from the streams, which get dries-up. Perfecting the techniques of induced breeding by hormones and stripping methods.

A fish sanctuary for protecting mahseer from attacks of illegal poachers and adored by many people. Several such sanctuaries exists in different parts of the country, most outstanding being at Hardwar and Rishikesh temples on the banks of Ganga river after its confluence with the Alakananda. In Madhya Pradesh too, there are sanctuaries on the Narmada at Kapileshwar, Mangalnath etc.

References

CIFRI. 1983. Final Report of All India Coordinated Research Project on Ecology and Fisheries of Freshwater reservoirs, Nagarjunasagar, 1971-81, CIFRI, Barrackpore, pp. 96.

Dasgupta, M. 1994. Mahseer of northeastern India-A review on biology. In: Mahseer the game fish (Ed. P. Nautiyal), Rachna Publication, Sri Nagar (Garhwal): B54-B66.

Day, F. 1878. Field notes of Sir Francis Day. Williams Clowes & Sons, London, pp. 44.

Desai, V. R. 1970. Studies on the fishery and biology of *Tor tor* (Hamilton) from river Narbada. I. Food and feeding habits. *J. Inland fish. Soc. India.* 2: 101-112.

Dubey, G. P. 1985. Conservation of dying king mahseer the mighty game fish, and its future role in reservoir fisheries. *Punjab Fish. Bull.* 9(1&2): 24-27.

Hora, S. L. 1940. The game fishes of India. IX. The mahseer or the large scaled barbels of India 3. The *Tor mahseer barbus* (Tor) Tor (Hamilton). *J. Bom. Nat. Hist. Soc.* 41(3): 518-525.

Jayaram, K. C. 1981. The freshwater fishes of India, Pakistan, Bangladesh, Burma and Sri Lanka. Zool. Surv. India, Calcutta, pp 475.

Nautiyal, P. and Lal, M. S. 1982. Recent Records of Garhwal mahseer (*Tor putitora*) with a note on its present status. *J. Bombay Nat. Hist. Soc.* 79(2): 430-431.

Sehgal, K. L. 1972. Coldwater fisheries and their development in India for sport and profit. Silver Jubilee CIFRI Souvenir, 125-131. Institue for Biological Exploration, San Francisco, 210.

Shaw, G. E. and Shebbear, E. O. 1937. Varieties of mahseer. *J. Darjeeling Nat. Hist. Soc.* 4: 22-25.

Surendranath, 1982. Extension of range of the putitor mahseer, *Tor putitora* (Ham.) cypriniformes cyprinidae (Barbinae) to Punch Valley (Jammu & Kashmir). *J. Bombay Nat. Hist. Soc.* 79(2): 430-431.

Chapter 6
Trout Culture

Trout is a well-known cold-water fish. Among cold-water fish, the trouts are esteemed owing to their sport value. They found mainly in the rivers, streams, lakes, brooks, and ponds of upland areas like Kashmir, Himachal Pradesh, Nilgiris, Uttar Pradesh, Kodai hills and Munnar high range.

Characteristic Features

Trouts have the following characteristic features:

(*a*) Streamlined body, (*b*) narrow gill openings and reduced gills, (*c*) adapted to highly oxygenated waters and freezing point temperature, (*d*) great powers of locomotion and clinging the burrowing habits, and (*e*) modified mouth and lips for rasping food particles with rocks, peebles etc.

The different species of trout are well suited for culture in India are the landlocked varities of brown trout (*Salmo trutta*), the rainbow trout (*Salmo gairdneri*) and the brook trout (*Salvelinus fontinalis*).

Salmo trutta fario (Linnaeus)

These fishes commonly known as brown trouts are the natives of the mountain waters of central and Western Europe. These were the first fishes reproduced and reared artificially in the country, but although they were introduced to the mountain waters of all the

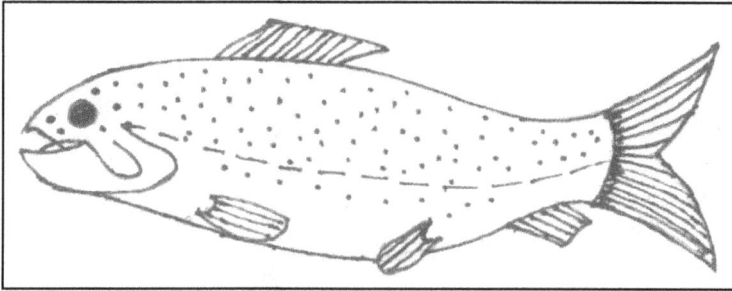

Figure 6.1: *Salmo trutta fario*

hills, they could establish themselves only in the streams and farms at Kashmir, and in river Beas in Punjab. The brown trout feed upon the crustaceans and large living prey at the bottom. It attains a maximum length of 18 inches, depending upon the availability of the natural food. During breeding season the fish swims upstream to spawn on gravel bedded shallows of fast current water. The brown trouts are timid and more demanding in comparison to other trouts, particularly when it comes to the quantity and the quality of waters.

Salmo gairdneri (Richardson)

These fishes commonly known as 'rainbow trouts' are the native of North American Pacific water and were imported to India in year 1907. Today they are one of the most successful trouts of Indian waters. They are considered hybrids of the several species of trouts from abroad. The rainbow trouts of Nilgiris and the Kodai hills constitute one of the important fisheries of Tamil Nadu. There are also six trout reservoirs, located at upper Bhavani, Mukurti, Prothumund, Parson's valley, Emerald and Avalanche in Nilgiris. In Kerala the fish has established itself in the streams at the high range of Travancore. The lake Devicolam, Eravikolam, Elephant and Luckham and the reservoir Kundally and Madupathy are the other important trout reservoirs of the state. The rainbow trout lakes and the farms of Kashmir and in various snow fed streams of Himachal Pradesh. These are, however, poorly represented from the Kumaun and Nepal hills. Rainbow trouts are best suited for cultivation because they adapt easily in comparison with the brown trouts. Moreover, they are less difficult to feed on artificial foods and can withstand the high temperature and oxygen depleted water. Their

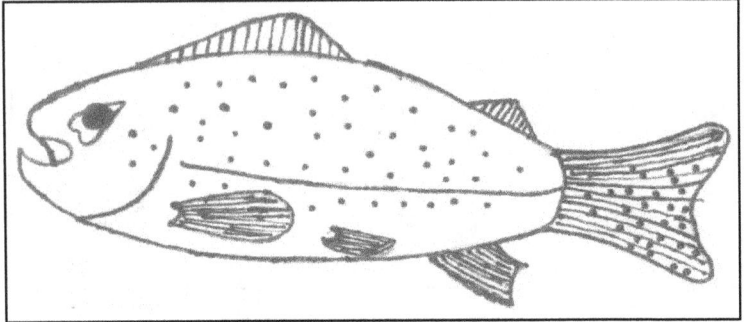

Figure 6.2: *Salmo gairdneri*

incubation period is shorter and the rate of development and the growth is faster. Upon being well fed, they attain a length of 40 to 50 cm in three years. They can withstand higher temperature up to 25°C. They have high stocking density, moderate growth and are more resistant to diseases.

Salvelinus fontinalis

These fishes commonly known as the brook trouts. It was transplanted in Indian waters in 1919 and later introduced from Kashmir to Kullu waters of Himachal Pradesh, but is now losing status of high demand because of poor adaptability. Their transplantation in Bhowate Naini Tal and several snow fed streams of Kumaun did not yield success in the long run. It is rather difficult to culture this fish because of poor tolerance to temperature (about 18°C). Secondly, this fish has the disadvantage of susceptibility to diseases (fin-rot, furunculosis) and also to pollutants or toxicants. The quality of its flesh is most palatable.

Trout Seed Resources

Trouts prefer gravelly substratum to safeguard their eggs and the fertilized eggs stick to gravel and debris, it is easy to collect the eggs in large numbers from such a substratum. The spawning season of trouts varies with temperature and the rainbow trout, *Salmo gairdneri gairdneri* spawns during September-February and the brown trout *S. trutta fario* during October-December.

Culture

The different phases of trout culture include (*a*) spawning and collection of eggs from brood fish; (*b*) incubation of eggs; (*c*) rearing of fry in nursery ponds, and (*d*) raising of fingerlings in production ponds, raceways, hatchery jars and drums provided with running water supply.

Egg Collection

The eggs and milt ooze freely from the vent of female and male brood fish when gently pressed. The eggs are collected in a black coloured enamels or plastic container to which the milt of the male is added. Immediately after adding the milt, the sex products are stirred with a quill for about 5-10 minutes when the fertilized eggs result. These eggs are adhesive in nature and are washed repeatedly with freshwater to remove the excess of milt and unwanted foreign particles. Finally, the eggs are allowed to remain in water for about one hour for hardening. It is known that about a teaspoonful of milt may be enough to fertilize eggs of two brood females of 500 g each.

The eggs and milt are viable for about three minutes. To ensure better survival rate, they may be collected in a small quantity of saline solution made of fresh water (10 liter), common salt (90 g), potassium chloride (2 g) and calcium chloride (3 g). The dead eggs could be distinguished from viable ones by keeping all eggs in 5 per cent glacial acetic acid for about 24 hours, when the dead eggs turn transparent while the viable ones, translucent. The fertilized eggs get a green tinge and are known as green, which are then transferred to hatcheries.

Incubation of Eggs

Eggs are incubated by keeping them either in concrete troughs with flat and horizontally arranged trays, incubators or jar hatcheries provided with circulating filtered and silt free freshwater.

Flat Trays and Troughs

Trays of wood, or aluminium painted with light coloured waterproof paint are invariably used to incubate the trout eggs. The bottom of these trays are provided with perforated zinc sheet, glass grills or mesh cloth for ensuring the passage of water through the different trays. The trays are horizontally arranged in the compartmental hatching trough, the size of which may vary from 180x30x10cm to 500x100x50cm. Provisions are made to pass the filtered water from a hatchery tank into the various compartments of the troughs and their trays fitted with inlet and outlet.

Incubators

These are of vertical-flow type. The trays are stacked one above the other. Each of them has two compartments, *viz.*, an upper egg basket and a lower perforated compartment on which the basket rests. The eggs are placed in the basket and water introduced in the bottom compartment of the tray to moist the eggs and facilitates hatching. Further, water is allowed to flow from tray to tray from the top and the excess water collected at the bottom is drained. The hatchlings are allowed to remain until they become swim-up fry stage. The tray units are then dismantled and the fry are transferred to larval tanks to make them swim freely and feed on supplemental diets.

Hatching Jar

The jar for use in the incubation of trout eggs has certain modifications. At the bottom of the jar and above the inlet, a galvanized screen of 0.5mm mesh with gravel bed is kept. This gravel bed acts as a filter to remove the unwanted particles. Above the filter, the eggs are placed for hatching. Water when passed into the jar through the inlet, upwells through the filter and eggs and drains through the outlet. Jar systems are used not only for the development of hatcheries but also to maintain them until they reach swim-up fry stage. The swim-up fry stage may be reached in a period of about three weeks after hatching in the jar systems.

Nursery Ponds

The nursery ponds may be concrete or stone-walled and take the size range, 2.5x1x0.75m- 9x1x0.75m. These are provided with suitable inlets and outlets for free water flow and the water flow inside the nursery pond may be fixed at 100 liter/min.

Rearing Pond and Raceways

In this, the advanced fry are grown to fingerlings or adult. A rearing pond may be a natural body of water or a raceway, which is merely a running water fishpond. A series of rectangular raceways ranging 20 to 100m² in size with a water depth of about 1.5m may be constructed either as all concrete structure or concrete sides with earthen bottom. The bottom of every raceway must be sloped so that it becomes shallower from inlet to outlet. Further, they are connected with one another and are supplied by the water of streams and nearby rivers. It is better, if the water –flow from these river systems into the fishponds is maintained at 50 litre/sec. It is also desirable to have a more or less constant water flow throughout the year. The inlet and outlet of each raceway are provided with zinc-plate screens. The inlet also has an extra-large meshed screen on the outside to prevent floating substances such as leaves and debris from entering. Depending on the water supply, oxygen content of the inflowing water and the artificial feeding, the sticking rate may be limited to produce from 5 to 10 kg/m². Growing trouts are fed with hearts and spleens obtained from slaughterhouse or with powdered marine and trash fish.

Besides easy management and harvesting, high production per unit area (200 kg/m²) is possible in raceways. Running water systems can also be constructed and used for carp culture, if suitable oxygen rich water sources are available.

Jar System

Owing to the fact that the rainbow trout is exceptionally tolerant of crowding, experiments have been conducted for growing trout from fry to adult stage in jar systems. Hatching jars of 7-liter capacity could be used; maintaining a water flow at 2.5 litre/min. Apart from high survival rate of trout fry, a production of 500 kg per cubic meter of water is possible.

Drums

Steel drums of 250-litter capacity with provisions as in jar hatchery could be used for growing trouts. It is possible to hold about 5,000-10,000 swim-up fry of rainbow trout weighing 20 kg in a single drum maintaining a water flow at 12 litre/min. To ensure maximum efficiency, the bottom of the drum should be conical. Fish in drums could be fed with dry feeds from the top. Though the flow pattern is from below, the feed sinks and no significant difference in food conversion has been noted between the drum and conventional unit. Rearing drums receiving a flow of 25 liter/min are self – cleaning. However, faecal materials and other debris from feed may accumulate along the bottom edge of the drum and hence, a monthly cleaning, using a flow of 50 liter/min is advisable.

Some of common diseases and their control: routs prefer to live in the clear running water with stream like conditions. However, they are not immune from infection by bacterial, fungal, parasite diseases. In Indian conditions five kinds of diseases in trouts have been identified. The types of diseases, main symptoms and treatment are given in Table 6.1.

Table 6.1

Disease	Symptoms	Remedial Measures
Whirling	Leaping and screw type of movement	The diseased trout should be isolated and eliminated
Pin head	Big head, and emaciated body	Mix calomel at 0.5 to 1 per cent with feed
Fin rot	Fin tips being eaten up	Given copper sulphate bath in the dose of 1: 2000
White spot	Appearance of white spots on fins, gills and body	Dip in formaldehyde solution (1 : 4000) for 15 minutes
Fungal attack	Presence of white patches on the body	Give treatment of malachite green solution (1 : 20,000) for 1 minute.

It has been marked by culture practices that the specimens of brown trout are more resistant to whirling diseases and those of rainbow trout are resistant to all kinds of diseases. However, the specimens of brook trout are less resistant to various kinds of diseases, pollutants as well as toxicants.

Chapter 7

Culture of Airbreathing Fishes

Culture of air breathing fishes is only occasionally practiced in India. On the contrary air breathing fishes are generally eliminated from the ponds because of their carnivorous feeding habit. However, if the farmers undertake the culture of these fishes, it would prove a more profitable venture, because of the following reasons:

1. They have a good market demand.
2. They are more nutritious and of high medicinal value.
3. They are highly recuperative
4. They grow faster and require little rearing
5. Culture involves low risk and simple management.
6. As most of them have air-breathing organs, their culture can be practiced in rejected foul water of shallow ponds, swamps and in other kinds of O_2 depleted waters.
7. When cage culture practice is adapted the income may increase manifold.

Presently farmed major carps, because of their popularity and availability at a relatively lower market price levels in general, do not fetch the level of economically viable market prices that would be comfortable. In this situation, the induction of air breathing fishes, which are known for their quality, taste and value features for farming in tanks and ponds, would zoom up the incomes of farmers. For achieving this reform of producing and bringing the coveted air breathing fishes closer to consumers the best possible agency that can be thought of is fish farmers development agency.

The *Clarias batrachus* and *Heteropneustes fossilis* among the catfishes, various species of channa among the murrels and *Anabas testudineus* constitute economically important fisheries.

These fishes being provided with air breathing organs, may be cultured in marshy and swampy places which constitute about 0.6 million hectares of unutilized waters of the country. Reclamation of these water potential is possible also by cage culture method of raising air breathing fishes.

Clarias batrachus (Linn.)

It is commonly called as 'magur' is in great demand in the north-eastern part of India particularly in West Bengal, Assam, Orissa and Bihar for its high nutritional value. In India, the states of Assam, Meghalaya, Tripura, Andhra Pradesh, Madhya Pradesh, Uttar Pradesh, Maharashtra, Tamil Nadu, Orissa and West Bengal support the most significant natural fishery of this air breathing fish. It contains higher percentage of protein and iron as compared to other edible species of fish. Its fat content is also very low and is, therefore easily digestible so that it is very useful during convalescence. With most other it is a delicacy because of the characteristic aroma and softness of its flesh. Fish fetches a higher price than the major carps.

Experiments based on modern cultural techniques have established beyond any shred of about that culture of magur is more profitable than that of any other cultivable species of fish. It is therefore; no surprise that popularity of magur culture from the point

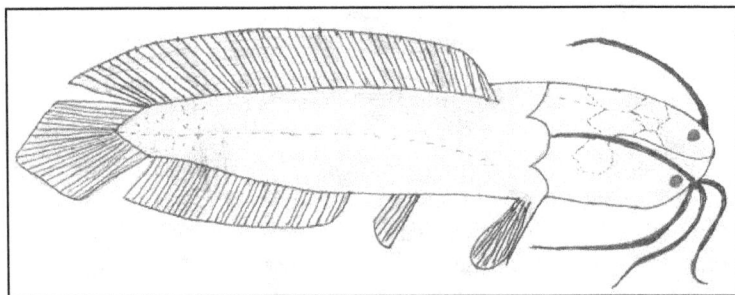

Figure 7.1: *Clarias batrachus*

of view of its food value and prospect of culture in confined boggy water body is gaining ground day by day. It is particularly so because in rural areas almost every household possesses a seasonal wastewater pond offering scope for culture of catfishes. The culture practices of this species have not received much attention, probably due to inadequate supplies of seed and proper feed.

Habit and Habitat

These fishes are not only able to thrive in water containing low dissolved oxygen but also they are extremely hardy with respect to all other environmental parameters and are suited to shallow and derelict waters as well as in saline impoundments having a salinity up to 10 ppt. It requires a relatively small area for culture and can be stocked at higher density than any other culturable species and even in perennial stretch of water as the culture periods is restricted to 5-6 months. The fish can live in water without oxygen and water with high content of CO_2 up to 71.45 ppm and thus can be cultured at a very high stocking density and give very high production per unit water area.

Food and Feeding Habit

Magur is basically carnivorous in habit. They prefer to eat decayed protein food and are considered as scavengers in fishpond. In cultural practices, they adopt themselves excellently to supplementary feeding with fish meal, oil cake, rice bran, earthworms, insect larvae, silk-worm pupae etc. In its adult stage it subsists on insect larvae, shrimps, worms, small fish and organic debris found in pond bottom. Young fry feed on protozoans, small crustaceans, rotifers and phytoplankton.

Spawning

In natural waters, the fish spawns during rainy season. For spawning, the fish swims to shallower regions of the already flooded ponds, swamps, streams, rice fields and other water bodies. Fish breeds once in a year. It attains maturity at the end of first year.

Fecundity

The fecundity of fish varies differently according to habits. Normally a fecundity of 3000-6000 numbers has been observed from fish ranging between 80 grams and 120 grams. Bigger fish above

130 grams are more fecund having fecundity to the tune of 10,000 nos or more.

Egg Laying

In nature, magur breeds in horizontal holes during the monsoon months from May to October. The fish migrate to adjoining inundated paddy fields and breeds in the grassy bottom nests prepared by them. In undulating terrain of Jhargram subdivision of Midnapore (West Bengal), magur has been found to breed in the nearby channels adjoining inundated paddy fields. Similar breeding behaviour has also been noticed in the deltaic region of Sunderbans. Abundant natural availability of magur seed is noticed from the paddy fields from the period of 'Bijya Dashami' to 'Bhatri Dwitiya'. The eggs are brown, gray black or dark greenish in colour, ranging from 1.2 mm to 1.8mm in size. The eggs are semiadhesive in nature, cling to the grassy and rocky substrate of the nests/holes. Fish breeds at a temperature ranging from 25°C to 30°C. Hatching take about 16-20 hours depending on the ambient temperature gradients.

Method of Culture

Culture of Clarias is more common inThailand. In the intensive culture of magur attempt have been made to obtain the production of 100t/ha/yr on experimental basis (Chaturvedi, 2002).

Magur fingerlings (80-100 nos/kg) are collected from natural resources like paddy fields, canals, swamps etc. Because of high mortality rate, larger ones are stocked by the farmer to harvest early crop. Normally magur is extensively cultured in West Bengal (as a companionate species) in conjunction with Indian major carps @ 7500-10,000 nos/ha.

Breeding Method and Hatching Technique

Despite large-scale natural collection of magur, there is a dearth of quality seed for intensive/semi intensive culture. Artificial propagation of magur was tried with success at Magra, Hooghly (West Bengal) by digging of horizontal holes (20-30 cm) inside the tanks below the water surface. The holes are filled with straws as substrate for adhesive eggs. One pair (male and female) of magur for each hole was liberated for breeding. No special care was taken for hatchlings. During the months of September to November, magur fingerlings were retrieved from tanks. The procedure as in vogue with, little modification in Thailand was followed.

Artificial Propagation of Magur

Pituitary gland extract or other hormonal preparations are given to male and female fishes @ 10-15mg/100gm of fishes (Chaturvedi, 2002). This fish can also be successfully induced to ovulate at 1ml ovatide per kg body weight ensuring high quality eggs and more normal larvae. Higher or lower doses affects the egg quality, led to spawning failure or low output of hatchlings (Sahoo *et al.*, 2005).

Experiments on the breeding of magur in controlled condition were conducted during the monsoon months of 1992 at the Freshwater Fisheries Research Station, Kulia, Kalyani, Nadia. Species of fully gravid male and female magur were collected from the specially managed brood fish pond where they were reared up with extreme care on a balanced daily ration of fish meal and rice bran. Pairs of male and female fish were administered 'ovaprim' composed of Salmon Gonadotropin RH and Domperiodone instead of fish pituitary gland extract. Ovaprim was used at a dose of 2.0 to 2.5 per kg body weight of male. The female ones attained the peak stage of ovulation after 15-16 hours of ovaprim administration when they were stripped off the ova in a tray. Simultaneously the grinded testis containing the milt of the injected males was thoroughly mixed with the stripped ova to obtain fertilization. After water hardening, the fertilized eggs were put into the 'glass jar hatchery, for further development. Hatching started after about 15 hours of fertilization at a temperature gradient of 28°C to 31°C.

Care of Fry and Fingerlings

Hatchlings are taken out from the 'glass gar hatchery' and kept in a cement cistern for subsequent rearing. Live zoo-plankters dominated by moina were fed to them for first three weeks to rear them up to 2-3 cm size for stocking in small pond for raising fingerlings for distribution to fish farmers for culture.

Harvesting

Growth of magur is very rapid in comparison with other walking catfish *i.e.*, *Heteropneustes fossilis*. It has been observed that the growth increment of fish averages 25-30 gm per month and to 120-150 gms in 5-6 months culture period at a stocking density of

50,000-60,000 nos/ha. The fish may be caught by scooping from the hiding-holes. The best way of harvesting of fish is by hand picking after complete dewatering.

Guidelines for Semi-Intensive Culture

In semi intensive culture practices, the following points should be carefully considered: 1) pretreatment of magur fingerlings with formalin bath (200ppm) for 20-25 minutes or with 0.3 per cent acriflavin for 5 minutes discarding the fish with loss of barbells. 2) Supplementary feed should contain 60 per cent of fish meal, 25 per cent of rice bran and 15 per cent of oil cake with vitamins and mineral mixture. Approximately 5 per cent of the body weight

Heteropneustes fossilis (Singhi or Stinging Catfish)

It is distributed in India, Sri Lanka, Thailand, Pakistan and Combodia. It inhabits the freshwater and swampy pools. Accessory respiratory organs are well developed. It adapts well to hypoxic conditions, withstands high stocking density and utilizes atmospheric oxygen when there is depletion in dissolved oxygen in the water body (Dehadrai *et al.*, 1985). Due to these organs fish can survive in muddy water or wet soil as well as out of water for considerable period. This fish is cultured along with other species in community ponds. Due to its high market value, rapid growth, tolerance to high stocking densities, utilization of atmospheric oxygen in oxygen depleted waters and low fat, high protein and iron content (Alok *et al.*, 1993; Ayyappan *et al.*, 2001), it can be cultured more intensively. Its protein content is higher but fat content is lower

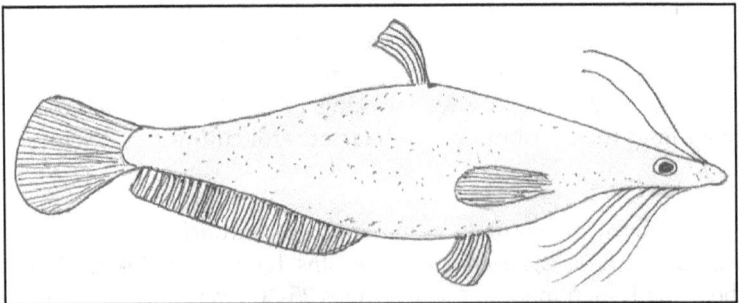

Figure 7.2: *Heteropneustes fossilis*

than that of major carps (Roy and Pal, 1986). Because of high haemoglobin content, it is recommended for the recuperation of patients.

The fish is suitable for culture in confined waters, swamps and marshes. It is bottom feeder and omnivorus. The fish feeds on protozoans, worms, crustaceans, insects and ostracods, algae and higher plants. It spawns throughout the year but chiefly in the monsoon.

It attains maturity in the second year and breeds once in a year during the monsoon season. In India considerable research work has been done on induced spawning of singhi. It could be now bred throughout the year in captive condition by hormonal manipulation (Munshi and Chowdhari 1996; Francis 1996; Alok *et al.*, 1998; Devi Pillai *et al.*, 2004). Induced breeding by hypophysation has also successfully attempted and spawning occurs about 8-10 hours after injection. In confined waters the majority of stock attains a length of 20cm at the end of the first year and grows to a maximum length of 45 cm. Fingerlings are cannibalistic.

These fishes can be cultured in confined water and are suitable for composite culture with other air breathing fishes.

Channa spp. (Murrels)

These fishes are commonly known as snakeheads. They occupy the top most rank among commercially important freshwater fish species. It is due to their unmatching taste, fewer intra muscular spines and medicinal qualities (Haniffa *et al.*, 2004). Murrels are air-breathers. They can survive in oxygen depleted water bodies and hence suitable for profitable culture in oxygen deficient areas like swamps and marshes. Murrels are esteemed as food and are known for their medicinal and recuperative values. The pricing of the murrel rages from Rs.250-300 at city wholesale fish market. Murrels are always sold live, since the prices are reduced by 30-40 per cent when dead. Murrels are excellent rounded meaty fish with single skeleton and minimum intramuscular bones. The firm and fleshy texture of flesh has a good meaty flavor. Murrels have an accumulation of fat deposits on the lining of the abdominal cavity and in gastrointestinal tract. Murrels have very thick belly muscle which makes it very good in steaking. Processing and value addition of murrels into various products *viz.*, boneless fillets, chunks, nuggets, keema and steaks

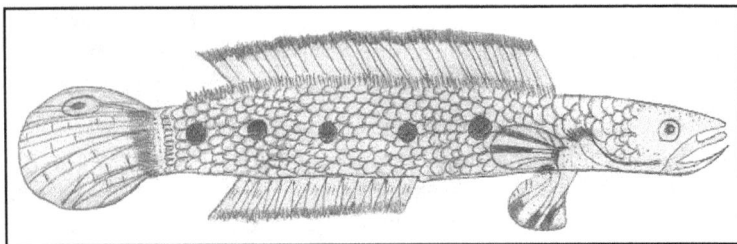

Figure 7.3: *Channa marulius*

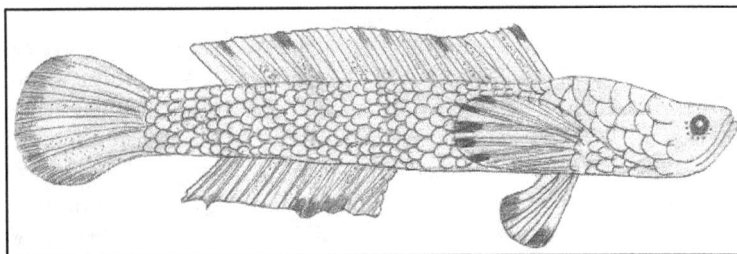

Figure 7.4: *Channa striatus*

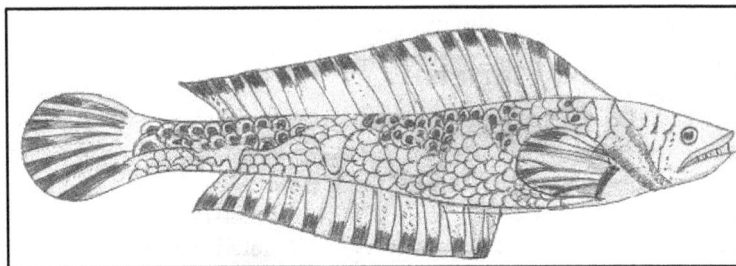

Figure 7.5: *Channa punctatus*

are described in detail by Sahu *et al.* (2011). Hundred grams of edible portion of murrel fish has moisture 78 gm, protein 16.2gm, fat 2.3gm, mineral 1.3gm, carbohydarate 2.2gm, energy 94 kcal, calcium 140mg, phosphrous 95 mg and iron 0.5mg (NIN, 1996).

Murrel population is dwindling in several parts of country. It is due to continuous collection of murrel fingerlings on the eve of Mrigasirakarthi day for dispensation to Asthma patients (fish medicine) that gather from all over the country in Hyderabad city.

Among the murrels, the stripped murrel, *Channa striatus*, the spotted murrel, *Channa punctatus*, the giant murrels, *Channa marulius* and *Channa micropletus* are widely preferred for human consumption in South-East Asia. Murrels have long been cultured in Thailand, Taiwan and Philippines.

Laximappa (2004) reported traditional system of culture of murrels in the irrigation tanks of Telangana and Royalaseema regions of Andhra Pradesh. The fingerlings of murrels are available in reservoirs, rivers, streams and other derelict water bodies. The demand for seed is mainly met from wild collections. Production of murrels under traditional composite culture system ranges from 50 to 150 kg/ha in 6-8 months culture period.

In West Bengal and Kerala there is a common practice of seasonal utilization of paddy fields for culture brackish water prawns and fishes. The culture of *Channa striatus* in paddy fields for 122days yielded about 13 kg of fish per hectare and at the same time increased the yield of paddy by 7 to 13 per cent.

Food and Feeding

C. striatus is a predator and it accepts feeds of animal origin. The baby larvae subsist on zooplankton while larger post larvae consume cladocerans and anisops, in addition to zooplankton. The large fry and fingerlings mainly prefer aquatic insects and shrimps while fry and fingerlings are active feeders. The intensity of feeding decreases with increase in size (Parameswaran, 1975) and with this, the fish develops piscivorous tendency and prefers as its food minnows, trash fish, frog tadpoles, annelids, gastropods etc in their food. The adults are predominantly piscivorous (Kuldip Kumar *et al.*, 2011).

C, marulius is predator. The feeding intensity increases during day time. The post larvae subsist on plankton comprising cladocerans, copepods, rotifers etc. With the increase in size, the larvae (25-50mm) feed mainly on small insects like anisops, water flea, hemipterans, zooplankters, annelids and small shrimps etc. As it grows, the fish prefers aquatic insects like coleopterans, nymphs of dragon fly, shrimps and annelids. With further increase in size, the fish feeds on small fishes. The adult murrels feeds on small fish, tadpoles, gastropods large aquatic insects etc (Kuldip Kumar *et al.*, 2011).

C. punctatus is also predator. It utilizes all niches of water boy during day time. The larva feeds upon plankton like crustaceans, and with the increase in size, the post larvae subsists upon copepods, crustancean larvae and rotifers. Later on, aqutic insects belonging to dipteral, hemiptera and odonata form the bulk of food (Kuldip Kumar *et al.*, 2011). Parameswaran (1975) reported small fingerlings feeding on aquatic insects, mainly anisop species, small shrimps, nymphs of dragon fly, annelids etc. The food of adult fish includes insects, shrimps, gastropods and fishes.

Maturity

C. marulius takes two years for maturity, whereas *C. striatus* and *C. punctatus* mature in the first year (Parameswaran, 1975). Murrels breed naturally during monsoons (Haniffa and Sridhar, 2002; Haniffa *et al.*, 2000; Marimuthu and Haniffa, 2004).

Sexual Diamorphism

In male brooder the vent appears pale in contrast to reddish vent in female. Females with soft and swollen bellies are generally selected for induced breeding. A gentle pressure on the belly portion for oozing of eggs will confirm the maturity.

Induced Breeding

Generally breeding set contains one female and two males. Brooders are induced to spawn by injecting pituitary or human chorionic gonadotropin or ovaprim or ovatide hormones, which are injected intramuscularly. After hormone injection the breeding sets are introduced into the breeding tanks. Aquatic macrophytes like Hydrilla and Eichornia are placed in the breeding tank to facilitate their hiding as well as for the egg attachment.

Spawning

Haniffa (2008) observed spawning behavior of murrels. Mating was preceded by elaborate courtship. Each female paired with single male. The active male chases the female and involved itself in the courtship. Male is generally observed hitting the female's snout and vent more frequently. The courtship continued till the release of gametes. At the culmination of courtship, the male bent its body close to the female and the brooders joined together; the male released its milt and the female its eggs and after that external fertilization takes place. The eggs are laid in area harboured with weeds. *C. marulius* construct a cup like nest in water. Nest building is not observed in *C. striatus* and *C. punctatus* (Haniffa, 2008; Haniffa *et al.*, 2004).

Development

Fertilization takes place within 24 hours after spawning. Ferilized eggs are usually buoyant and adhere to each other forming an egg mass containing 5000 to 10,000 eggs in the case of *C. striatus* and 3000-6000 eggs in *C. punctatus*. Eggs of *C. marulius* are green yellow, measuring 1.5-2.0mm diameter. In case of *C. punctatus* eggs are spherical and non-adhesive. The rate of fertilization varies from 60-80 per cent. Hatching takes place 24-30 hours after fertilization. The hatchlings of *C. striatus* measure from 2.8-3.2mm in length, while the hatchlings of *C. punctatus* measure from 2.7-2.9cm. *C. striatus* hatchling was 2.80 mm in length and 10 days old post larva at about 6.2mm in length. Fry is about 7.3 mm in length. Generally from a single spawning 3000 to 9000 hatchlings are obtained. The male moves around the hatchlings and ventilates them with pectoral fins.

Artificial Fertilization

If spawning behavior was not noticed within 24 hours after injection of the hormone and when the induced breeding attempt failed, then the artificial fertilization is done by stripping method. Stripping is done by gently pressing the abdominal portion with the thumb from the pectoral fin towards the vent of female. Eggs oozes out. These eggs are collected in a plastic bowl. As the males never ooze out milt while stripping, the sperm can only be obtained by sacrificing the males. The testes need to be rapidly out into small pieces using a pair of scissors and finally the testes are ground using a glass homogenizer using 0.9 per cent physiological saline solution. Sperms are directly added into bowl containing the eggs. Eggs are allowed to fertilize by adding an equal volume of water and by gently shaking the bowl. After about one minute fertilization takes place. After fertilization the eggs are washed in distilled water to remove excess milt. The hatchlings are reared in cement tanks.

Larval Rearing

Absorption of yolk sac takes place within 3 to 4 days and by that time the mouth is fully forms. The post larvae began to feed on exogenous feed. The post larvae are reared in the cement tanks. During the early post larvae stage small plankters are provided. After 15 days, during the late post larvae stage, daphnia, moina along with chicken intestine paste is used as food.

Fry Rearing

Fry is stocked @ 300-500/m2 in masonry tanks (3mx1mx1m) and fed on chicken intestine along with trash fish and tapioca flour as semi-moist dough. The fry were given the semi moist feed twice daily.

Anabas testudineus (Climbing perch or Koi)

It is distributed in Indonesia, India, Sri Lanka, Pakistan and Burma. The colour is rifle green, becoming lightest on the abdomen. *A. testudineus* grows to a maximum length of 23 cm.

It is popularly known as Koi in most parts of India is a highly demanded fish. It is an important air breathing fish, which can be considered for culture in the areas with low dissolved oxygen. It is also popular for their lean meat, which contain easily digestible protein and fat of very low melting point and many essential amino acids making them ideal food. So, anabas enjoys a good market demand in India particularly in the north-east region.

This fish feeds on small fishes, crustaceans and insects. *A. testudineus* can also tolerate the brackish water environment. These breed in confined waters like ponds and reservoirs. The rice fields

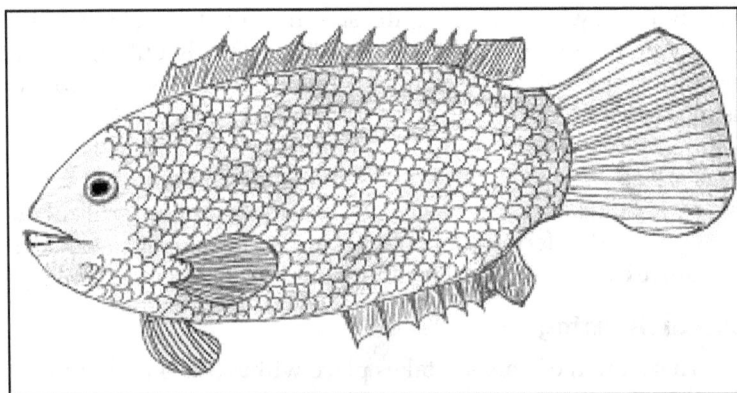

Figure 7.6: *Anabas testudineus*

are suitable for cultivation. They attain sexual maturity when about 20mm long and about 6 months old. They breed during monsoon period. The eggs are minute, yellowish or whitish and floating freely on the surface until they hatch. The embryo hatches out after 24 hours at a temperature of 28.5 °C. The hatchlings are small and tender. After 8 mm stage, the spawn readily accepts zooplankton and is highly cannibalistic till it reaches a size of 22mm when the fry acquires the adult characters.

Now anabas is a threatened species due to over exploitation and conservation measure is very much needed. Induce mass breeding and culture of this species may be ensuring to sum the fish in nature. This fish has a tendency to walk away from the pond during rains. Above all, the main constraints of culture of this fish are non availability of seed from natural resources as well as artificial propagation. So it is essential to develop some of the standard breeding methods to overcome the cultural constraints.

Many scientists have made several attempts for artificial breeding of anabas with varying degrees of success. Bangladesh has succeeded in developing the technology for artificial breeding under controlled conditions. However partial success has been made in India and yet to be standardized the technology. A brief description of breeding behavior has been discussed here.

Maturity occurs at the age of one when the fishes reach a size of 10-12cm in total length. The sexual dimorphism in *A. testudineus* is more apparent during breeding season. The mature male acquires a reddish hue on the body, particularly on the pectoral and ventral fins. The female shows only a faint reddish color. Further in the male a distinct diamond shaped black spot appears in the caudal peduncle. In the female this black spot is oblong and somewhat diffused. Moreover, the female is contrast to the male, has a prominently bulging abdomen. The ventral distance between the bases of the two pectoral fins in the female is significantly greater than the male. In the breeding season, the female exhibits a prominent bulge at the vent, resembling the genital papillae while in the male this structure is absent. Mature male oozed out white milt and mature female oozed out ova even at a gentle pressure at the abdomen during breeding season.

In nature the eggs are scattered in open water at the onset of the rains without any nest. The male wraps itself in the female body, fertilizing the eggs as they are laid. Each time 200 colorless eggs are released until about 5000 numbers are laid. The fecundity varies from 5000-35000 numbers. The eggs rise to the surface and float. The eggs hatch in 24 hrs and the fry are about 2-3 mm long. They are free swimming within two days of hatching.

In case of artificial breeding with pituitary or synthetic hormone, a single dose of injection for both the male and female spawning actively and courtship behaviour starts after 6 hrs of injection. The water temperature to be maintained at 28° C + 1°C. Fertilized eggs float in the surface of water. It takes 18-20 hrs for hatching after spawning and newly hatches larvae measures 1.8-2.0 mm in length without any movement. Yolk sac completely absorbs on third day after hatching and settles at the bottom. Plankton and Artemia are supplied as artificial feed for those fries up to 20 days. The survivability is about 75 per cent.

It can be cultured singly or in combination with magur and singhi. It can also be grown in combination with carp fingerlings of over 10 cm sizes. This practice can be employed in order to utilize the insect fauna as well as in respect of the role of fish as a biological check on small insects in water.

At Kalayani (West Bengal) a pond of 0.02 hectare stocked at a rate equivalent to 125,000/hectare with anabas fingerlings gave a production to 702 kg/hectare in 11 months despite a low survival of 21.8 per cent due to breeding migration. Cultutre of anabas in barrel-shaped bamboo cages resulted a production of 12.7 kg/cage in five months.

These may be cultured alone but often in association with channa and clarias in ponds and rice fields. They attain a length of about 120mm at the end of first year and about 200mm at the end of second year.

References

Alok, D., Krishnan, T., Talwar, G. D. and Garg, L. C. 1993. Induced spawning of catfish, *Heteropneustes fossilis* (Bloch), using D-Lys[6] Salmon gonadotropin-releasing hormone analog. Aquaculture. 115: 159-167.

Alok, D., Krishnan, T., Talwar, G. D. and Garg, L. C. 1998. Multiple induced spawning of the Indian catfish *Heteropneustes fossilis*

(Bloch) within a prolonged spawning season. World Aqua. Soc. 29(2): 252-258.

Ayyappan, S., Raizada S. and Reddy, A. K. 2001. Captive breeding and culture of new species of aquaculture. In: Jhingran, A. G., Lal, K. K., and Basheer, V. S. (Eds.), Captive breeding for aquaculture and fish germplasm conservation. NBFGR-NATP Publication-3. Key Paper No. 1. National Bureau of Fish Genetic Research, Lucknow, Uttar Pradesh, India.

Chakraborti, P., Maitra, G. and Samir Bhattacharya. 1986. Binding of thyroid hormone to isolated ovarian nuclei from a freshwater perch, *Anabas testudineus*. *General and Comparative Endocrinology*. 62(2): P239-246.

Chaturvedi, C. S. 2002. Biology, culture and breeding technique of Indian magur *Clarias batrcahus*. In: Carp and catfish breeding and culture (Complied by Langer, R. K., Dube Kiran and Reddy, A. K.), Central Institute of Fisheries Education, Mumbai, pp. 74-79.

Dehadrai, P. V., Yusuf, K. M. and Das, R. K. 1985. Package of practices for increasing production of air breathing fishes, p. 1-4. In: Auaculture Extension Manual, Information and Extension Division of CIFRI (ICAR), India, New Series No. 3.

Devika Pillai, Alok, D. and Garg, L. C. 2004. Induced ovulation in catfish *Heteropneustes fossilis* (Bloch) with three native GnRH peptides. *Fishery Technology*. 41 (1): 5-8.

Francis, T. T. 1996. Studies on the effect of pituitary hormones and feeds on the reproduction of *Heteropneustes fossilis* (Bloch). Ph. D. thesis, Fisheries College and Research Institute, Tamil Nadu Veterinary and Animal Sciences University, Tuticorin, India, Thesis abstract, Naga. 19(3): 33.

Haniffa, M. A. 2008. Mass seed production of murrels is an instant need. *Fishing Chimes*. 28 (1): 111-115.

Haniffa, M. A., Marimuthu, K., Nagrajan, M., Jesu Arokiaraj, A. and Kumar, D. 2004. Breeding behavior and parental care of the induced bred spotted murrel *Channa punctatus* under captivity. *Current Science*. 86 (10): 1375-1376.

Haniffa, M. A., Merlin Rose T. and Shaik Mohamed J. 2000. Induced spawning of the stripped murrel *Channa striatus* using pituitary extracts H. C. G. LHRHa and Ovaprim. *Actaicht. Piscat.* 30(1): 53-60.

Haniffa, M. A. and Sridhar, S. 2002. Induced spawning of the spotted murrel *Channa punctatus* and the catfish *Heteropneustes fossilis* using ovaprim and human chorionic gonadotropin (HCG). *Veterinarski Archiv.* 72(1): 51-56.

Kuldeep Kumar, Eknath, A. E., Sahu, A. K., Mohanty, U. L., Rajesh Kumar, Sahoo Minakshi andNoor Jahan. 2011. Snakeheads: Challenging fish for diversification of fish farming. *Fishing Chimes.* 31(1): 110-113.

Laxmappa, B. 2004. Status of murrel farming in Andhra Pradesh. *Fishing Chimes.* 23 (12): 60-61.

Marimuthu, K. and Haniffa, M. A. 2004. Seed production and culture of snakehead. *Infofish International.* 2: 16-18.

Munshi, D. J. S. and Chowdhari, S. 1996. Ecology of *Heteropneustes fossilis* (Bloch) an air breathing catfish of South-East Asia. Narendra Publication House, New Delhi, 165pp.

N. I. N. 1996. Nutritive value of Indian foods, National Institute of Nturition, 1-156pp.

Paramesvaran, S. 1975. Investigation on the biology of some fishes of the genus *Charma gronovius*, Ph. D. Thesis, Magadh University, Bodh Gaya, 299.

Roy, S. and Pal, B. C. 1986. Quantitative and qualitative analysis of spawning behavior of *Heteropneustes fossilis* (Bloch) (Siluridae) in laboratory aquaria. *J. Fish. Biol.* 28: 247-254.

Sahoo, S. K., Giri, S. S. and Sahu, A. K. 2005. Effect on breeding performance and egg quality of *Clarias batrachus* (Linn.) at various doses of ovatide during spawning induction. *Asian Fisheries Sciecnes.* 18: 77-83.

Sahu, B. B., Kuldeep Kumar, Sahu, A. K., Mohanty, U. L., Rajesh Kumar, Meenakshi Sahoo, Noor Jahan and Ambekar, E. E. 2011. Processing and value addition to murrels in value chain. *Fishing Chimes.* 31(1): 106-108.

Sarkar, U. K., Deepak, P. K., Kapoor, D., Negi, R. S., Paul, S. K. and Sreeprakash Singh. 2005. Captive breeding of climbing perch *Anabas testudineus* (Bloch, 1792) with Wova-FH for conservation and aquaculture. In *Aquaculture Research.* 36(10): 941-945.

Chapter 8

Cage Culture

Cage culture is a type of enclosure culture, involve holding organisms captive within an enclosed space while maintaining a free exchange of water. A cage is totally enclosed on all, or all but the top, sides by mesh or netting. Cages are smaller, and typically have a surface area somewhere between 1m^2 and 1,000m^2. Because of their small size, cages are better suited to intensive culture methods.

The origins of cage culture are a little vague. It can be assumed that at the beginning fishermen may have used the cages as holding structures to store the captured fish until they are sent to the market. The first cages which were used for producing fish were developed in Southeast Asia around the end of the 19th century. Wood or bamboos were used to construct these ancient cages and the fish were fed by trash fish and food scraps. In 1950s modern cage culture began with the initiation of production of synthetic materials for cage construction. During last few years, the practice of cage culture in inland waters has spread throughout the world to more than 35 countries inn Europe, Asia, Africa and America. By an estimate more than 70 species of freshwater fish have been experimentally grown in cages.

Site Selection

An ideal cage site must have good water quality, which means that it should not only be uncontaminated by toxic industrial pollutants. The important water quality parameters like pH, temperature, and oxygen should be carefully considered. The actual supply of dissolved oxygen to the caged fish is dependent not only upon the dissolved oxygen concentration in water, but also upon water exchange through the cage. If the supply of oxygen to fish deviates from the ideal, than feeding, food-conversion, growth and health can be adversely affected. Normally cage fish farmers should avoid higher turbid sites. The industrial effluents could damage the cage culture, cultured fish, its food and accumulated toxicants in fish may be harmful to humans. These risks may be significantly reduced by placing cages far away from industrial areas. Cage culture should be avoided in disease risk areas, usually organically polluted water bodies seem to harbour more disease agents than unpolluted waterbodies. Good water exchange or flushing at the site is essential and it minimizes the buildup of waste metabolites inside the cage.

The nature of substrate at a site has a considerable influence on the choice of cage design viz; fixed cages or floating cages. The cages should be sited in sufficient depth to maximize the exchange of water, and yet to keep the bottom of the cages well clear of the substrate. The site of cage to be installed should preferably be in a protected or sheltered zone essentially free from direct action of wind, and impact of waves.

Security is a main problem for cage fish farmers in many parts of the world, since cages are often located in water bodies, which are publicly owned or have unrestricted access and are thus vulnerable to poachers and vandals. The cages should be located at sites which are less disturbed and where these can be monitored once in each day.

Water Quality Monitoring

The success of cage culture depends on maintaining good water quality around the fish cages. Watershed should not be a highly erodable one or one that allows the accumulation of large amounts of organic debris. The water level should not fluctuate greatly. Problems with aquatic weeds, overpopulations of wild fish and oxygen depletion problems should not be present. The desirable range of early morning pH for fish production is from 6.5 to 9. Dissolved oxygen concentration and its availability are critical to the health and survival of caged fish. In general, warm water species such as catfish and tilapia need a dissolved oxygen concentration of more than 4 mg/l to maintain good health. Dissolved Oxygen levels below 3 mg/l can stress fish. If this level goes below 2 mg/l can increase the mortality of fish (Liyanage *et al.*).

Fish Species Raised in Cages

Among the species, *Pangasius sutchi, P. micronemus, Cyprinus carpio, Salmo salar, Salmo gairdneri,* and *Tilapia nilotica* are extensively cultured in cages in other countries. The fishes commonly used for cage culture in Japan are yellow-tail (*Seriola quinqueradiata*) and sea bream (*Chrysophrys major*). Culture of carp (*Cyprinus carpio*) and other varieties of freshwater fish is also practiced in cages on a large scale in Japan. In Indian freshwaters, the fish species raised in cages are *Catla catla, Labeo rohita,* and *Cirrhiuns mrigala, Cyprinus carpio, Hypophthalmichthys molitrix, Ctenopharyngodon idella, Mystus seenghala, Ompak bimaculatus* and *Heteropnestus fossilis* have given encouraging results. The above-mentioned fishes offer excellent potential for this type of culture. They grow rapidly and have high survival rate. Few workers mentioned the possibility of growing high priced catfishes like *Clarias batrachus* to market size in cages.

Cage Design

The design of cages varies with the behaviour of fishes to be reared. The pelagic fishes have a general tendency to swim swiftly often in an orbital motion near the surface of water. For such fishes circular, hexagonal or spherical end edged cage may be found suitable. Cages may prove more useful for demersal fishes, which prefer to move in midwater and deep water. The dimensions of culture cages vary considerably (from $1m^3$ to more than $100m^3$) according to the kind of construction material, the type of culture, and the local conditions. In the United States, for commercial production of catfish $1m^3$ cage has been used to facilitate handling and harvest. In the Netherlands, the intensive production of rainbow trout and carp is carried out with great success in net cages of only $6.5m^3$, grouped together in fours to make production units of $26m^3$ each. In India, small sized cages in the range of $4.5 m^3$ to $6m^3$ have been used.

The size depends on the strength and durability of the materials to be used in building the cage, the species to be reared in the cage and available management facilities. With regard to cage construction materials, bamboo and wood are used in south East

Figure 8.1: Cage Culture in Lake

Asia for traditional cage farming. But in intensive culture practices, synthetic netting cages are used in Japan, rigid metals and plastic meshed materials in USA. Generally cages are available in polyethylene, polyamide, polyester and dynema. Polyamide netting of different kinds such as dedron, nylon, capron or perlon are less expensive. But wire mesh is used more owing to its advantage of easy cleaning. Despite the many advantages of using other material (stronger, longer life etc) bamboo is still the most commonly used construction material for cage frames in South-East Asia. It is due to its low price and easy availability.

Carrying Capacity

Carrying capacity of the site (*i. e.*, the maximum level of production that a site might be expected to sustain) is an important aspect. Intensive cage fish culture results in the production of wastes, which can stimulate productivity and alter biotic and abiotic characteristics of the water body, while less intensive methods can result in overgrazing of algae and a fall in productivity. A serious deterioration in water quality will stress or even cause mortalities amongst stocks and can encourage disease organisms to thrive, while overgrazing of algae can result in poorer growth and an increasing reliance on supplementary feedstuffs. Hence profitability or even viability may be affected. Hence it is important for all concerned with cage fish farming development that an accurate evaluation of the suitable levels of production at a site is made.

Fish Production

The channel catfish, *Lactarius punctatus* in USA yielded production of 20-35 kg/m^{-3} In Africa, *tilapia* yielded 17 kg/m^{-3} and trout produced 15 kg/m^{-3} In case of carp culture the yield in Japan, with intensive artificial feeding and after 5 months, the average yield is 30-40 kg of fish/m^2. It amounts to 300,000 to400, 000 kg per hectare. In term of profit margin it is about 10-12 times more than with the traditional pond culture.

Indepesca Aquaculture Pvt. Ltd., Mumbai has created a landmark by establishing and demonstrating a pilot scale cage culture project comprising 12 cages in the Dhasai reservoir of Thane district, Maharashtra. These cages were set up in Dhasai reservoir in September 2009 and stocked with seed of Pangasius in December 2009. The dimension of each of the cages is 4x6x3m and 190 fish were stocked per cubic meter volume of water (13800 fish per cage) to provide 13000 kg of biomass of fish per cage at harvest. The initial average stocking size of the seed was 25 g each. Fish were fed using the soy-based extruded floating fish feed with the satiation feeding technique recommended by American Soybean Association-International Marketing (ASA-IM). The feed was manufactured by the Indian Broiler Group, Chattisgarh which contained 28 per cent protein and 3 per cent fat. Young ones were stocked at a 25 g of average weight each, after weaning them onto formulated feed in a nursery at Nagpur Fish Farm. The satiation was re-worked after every 10 days to ensure that fish received adequate feed in relation to their growth. The cages were regularly monitored and cleaned to remove fouling organisms and algae and to ensure an efficient flow of water through the system. The cages yielded a production of about 13mt/cage which translated to about 180kg fish per cum and their body weight averaged at 1 kg per fish. The FCR obtained in the system was 1:1:5. They were sold at a farm gate price of Rs.44 to be taken to local Mumbai market. Based on this pilot scale project, the Indepesca Group feels that it is a viable venture if the number of cages could be expanded. The management has already taken a decision to expand the number of cages to 60 by December 2010 and expand commercial operations.

Rajiv Gandhi Centre for Aquaculture (RGCA) has demonstrated technical viability of seabass culture in cages set up in ponds, both in nursery and grow out culture phases by providing necessary infrastructure. RGCA had taken up post-nursery rearing of hatchery raised early fingerlings of 50mm size with weight of 1.7 to 2.0 gm each. Fingerlings were stocked in nylon knotless cages of 2mx2mx1.3m, having an area of 5 cu. m. About 27,000 nos of fingerlings were stocked in 4 cages initially @ 1400-1500 nos/sq. m. Accordingly 7000 nos of fingerlings were stocked in each cage initially, by maintaining the average biomass of 2.5-3.0 kg/cu. m. The fishes were fed with 1.2mm, 1.5mm and 2.0mm extruded pellet feed (slow sinking type) @ 6-8 per cent of the body weight. The grading frequency is 8-10 days. The fingerlings were reared for a period of 30 days. The grading of fry and fingerlings involves the collection of all the fishes from net cages and placing them in grading box. The smaller fishes are able to swim out of the grader through the narrow opening of the movable pipes, and larger fishes are left behind. These are transferred into another tank. Three or four grades are obtained from each grading depending upon the grading and frequency of grading. The survival during nursery phase was 87.2 per cent *i.e.,* 24,000 nos. The size of the fishes harvested after post nursery rearing was 80-100mm within the weight range of 10-15 gm each. The FCR during nursery phase was 1.23:1. The biomass maintained at the end of the nursery phase was in the range of 5.50-6.50 kg/cu. m. The number of cages maintained at the end of the nursery phase was 10. The growth of each of the fingerlings per day was in the range of 0.32 -0.45 gm. The fishes were harvested after a farming period of 217 days. The growth of fish in cages was in the range of 600-720 gm each. The successful demonstration of farming of seabass in pond cages by RGCA has proved that the cage farming is a technically viable project having 87.5 per cent survival in grow-out phase with a computed 12.4 tonnes of production per hectare of cage area. Total number of cages harvested were 126 with average biomass of 98.4 kg/cage.

In Kabini reservoir (Karnataka) the seed of catla and rohu were reared separately to advanced fingerlings stage in six cages of 3m x 3mx2m (1.75 m below water level). The total volume of each cage was 15.75 m³. The stocking density is 100 no/m³ and 175 np/m². The total seed stocked per cage was 1575 nos. The frame was fabricated using bamboo poles and used oil drums were recycled for floatation. The cage bag was stitched using HDPE fabric of 12 mesh/inch (2mm mesh bar). Each frame accommodated three cages. The feed consisting of rice bran and ground nut oil cake in the ratio of 1:1 was given three times a day at 10 per cent of the body weight. Catla had grown from an initial size of 4.0 cm (1.4 g) to 7.0 cm (4.3 g) and rohu from 5.2 cm (1.4 g) to 6.6 cm (3.3 g) in 60 days.

Cage culture experiments carried out by CIFRI at Allahabad recorded a production of 16 kg/m² in case of *C. mrigala* while its polyculture with other two species *viz. Catla catla* and *Labeo rohita* gave only 2 kg/m² production. Culture of *Anabus testudinues, Channa punctatus* and *C. striatus* in Bangalore has given production range of 0.3 to 1.75 kg/m²/month. A production rate of 0.8 kg/m²/month with catla and mrigal mix was achieved in Jari tank at Allahabad. On the other hand with common carp a production of 1.54 kg/m²/month was achieved in Sankey tank at Bangalore. The institute also successfully raised fingerlings of common carp and *Catla catla* in cages. The stocked fry recorded a survival rate of 90-97.5 per cent and attained fingerlings size (100mm) in about 2 months.

In a series of net cages installed in Ennore estuary postlarvae of *Penaeus indicus* gave production range of 1,250 to 2,880 kg/ha while *P. monodon* recorded a production of 1,450 kg/ha.

Preliminary trials by CIFRI to culture fish and prawn in cages installed in the beels of West Bengal produced encouraging results. Nelton screen of 5mm mesh size of black colour was selected for making the cage. The desired shape of the screen was set by sewing the outer edge with nylon thread for giving the shape of a box. The box was further strengthened by making the frame with acrylic pipes. The cage erected in Mathura beel was 3mx2mx2m. The cage was positioned in place with the help of 4 bamboo poles in the open area of the beel with provision of proper water circulation and protection from strong currents and higher waves. The cage was fixed at a height of 0.4m from above the bottom of the beel. Catla and rohu fingerlings of 100-180 cm size were stocked at a density of 10,000 nos/ha. Fishes were fed regularly with pelleted feed containing 25 per cent protein @ 2 per cent of the body weight. The growth of catla and rohu has been recorded on an average 300 g and 183 g respectively. The production obtained from the cage estimated at 11 kg/12m³/130 days.

The CMFRI successfully harvested farmed spiny lobsters from the floating cage moored in the Vizhinijam bay, Tiruvananthapuram district of Kerala. At Vizhinijam, 1200 juvenile lobsters (*P. homarus*) weighing between 70 and 95 gm were stocked in 15 July 2008 in a 5 m diameter floating cage with a 6m cylindrical HDPE netting protected outside by another net. The cage was moored at a depth of 8 m in Vizhinijam bay. Lobsters were fed daily ad libitum with live mussels collected from nearby rocks. Growth of lobsters and environmental parameters around the cages were regularly monitored to ensure good growing conditions. Lobsters at harvest weighed an average 250g each after four and half months of on growing and 85 per cent of lobsters stocked were retrieved. The harvested lobsters were as healthy as wild caught lobsters and had good colouration. With a price tag of Rs.1000/kg, the farming operation has turned out to be an economically successful venture with a net profit of Rs. 1.5 lakhs.

The cage culture experiments were carried out under an approved research project of CIFE, at Walwan reservoir of Pune district (Maharashtra), Powai lake (Mumbai), Halali reservoir (Madhya Pradesh) and Gobindsagar reservoir (Himachal Pradesh). The fishes cultured were major carps and mahseer.

Indian major carp fingerling were raised in cages in Jaismand lake of Rajasthan (Sharma *et al.*, 2009). The performance of catla and rohu were much better than mrigal n the two experimental cages. In all 1,800 fish (early fingerlings) were released in a ratio of 1:6:3 catla, rohu and mrigal in each of the two cages set up. At the time of harvest in the first round of rearing, a survival of 65 per cent was noticed. Relatively, the performance of catla was the best followed by rohu. Mrigal showed relatively poor growth which may be due to non availability of their choice food in the cage environment. For ensuring good survival, a second batch of fingerlings (112 nos) was introduced in the first cage and 113 in the second cage. These comprised rohu, catla and mrigal in the same ratio of 1:6:3 as before. The rearing period was from 15 January 2004 to 17 March 2004 where in main emphasis was on checking the rate of survival. After the rearing period of about two months, the fish were harvested and released in the lake. At this time the survival was excellent *i.e.*, out of

225 advanced fingerlings introduced in two cages the recovery was 99.5 per cent. The fish were healthy in them without symptoms of any disease. From a comparison of the first and second rearings, it was evident that, before releasing in cage, the fingerlings should be acclimatized to the cage conditions to reduce mortality. It was also seen that sizable quantity of periphyton developed on the webbing of cage would serve as a main diet of rohu.

Advantages of Cage Culture

1. Cages improve the quality of the harvested fish
2. Cage farming saves the time and labour. Two persons can harvest the fish in a cage without difficulty.
3. Sick or diseased fishes in cages can be segregated conveniently for observation or treatment.
4. Recording of the various parameters, such as rate of feeding, growth rate and observations in respect of diseases would be easier in the cage farming system.
5. Free exchange of water is possible in cages.
6. Several units of cages could be installed in a water body for gainful employment and income.
7. Large water bodies could be utilized better for fish culture
8. Oxygen depletion cannot be found in cages
9. Fishes grown in cage culture would not develop muddy flavour.
10. Fishes grown in cage provides maximum value due to the high quality and harvest of fishes can be done according to market trend.

Disadvantages of Cage Culture

1. Stock is vulnerable to external water quality problems such as algal blooms and low oxygen levels
2. Fish growth is significantly influenced by ambient water temperatures
3. Feed must be nutritionally complete and kept fresh
4. The incidence of disease can be high and diseases may spread rapidly
5. Poaching is a potential problem

References

CMFRI. 2009. Cage farming and harvesting of spiny lobster: A success story of CMFRI, Vizhinjam Bay, Kerala, 2 June 2009, pp. 38-39.

Sharma, L. L., Durga, I. A., Bhatnagar, Manoj and Sharma, S. K. 2009. Ranching of cage-raised IMC fingerlings in Jaismand lake, Rajasthan. *Fishing Chimes*. 29 (1): 91-94.

Thampi Sam Raj, Y. C. and Pandiarajan, S. 2010. Grow-out farming of seabass in pond cages. *Fishing Chimes*. 30 (1): 89-91.

Vijay Anand, P. E. 2010. Cage farming of Pangasus (*Pangasius hypopthalmus*): Demonstration in a minor irrigation reservoir, Maharashtra. *Fishing Chimes*. 30 (9): 8-9.

Liyanage, N. P. P., Ruwanpathirana, S. M. and Jayamanne, S. C. 2009. Cage culture of freshwater fishes in reservoirs. Training manual for Kattakaduwa fishing community, NARA-AIDA Project 2006-2009.

Chapter 9
Pen Culture

Pen culture is a type of enclosure culture; involve holding organisms captive within an enclosed space while maintaining a free exchange of water. In pen the bottom of the enclosure is formed by the lake/beel bottom. Pen enclosures tend to be bigger in size ranging from 0.1 to 1000 ha. The origin of pen culture is obscure, but it also seems to have begun in Asia. As per reports, pen culture originated in the Inland sea area of Japan in the early 1920s. It was adopted by the People's Republic of China in the early 1950s for rearing carps in freshwater lakes, and was introduced to Laguna de Bay and Sam Pablo lakes in the Philippines in order to rear the milkfish *Chanos chanos*.

Selection of Site

Selection of site for pen enclosure is of paramount importance as this aspect largely decides the economic viability and success of the farming system. Hence before undertaking the construction of pen enclosure, it is essential to undertake detailed engineering studies at the site on the following aspects: 1) the normal depth of water in the watershed. Whether the lake and surrounding areas get inundated by flood, flash flood, storm surge etc, and if so, the maximum depth of inundation, 2) shape, size and nature of the watershed, whether running or standing, 3) infestation of aquatic vegetation, if any, 4) characteristics of the catchment area, vegetation cover etc., 5) river or stream discharge, high flood level in the river, if there is any adjoining watershed, 6) Topographical; features of the shore, g) type of soil, if sandy, loamy, clayey or silty 7) availability of cultivable species 8) Pollution problem, if any 9) accessibility 10) intensity and duration of rainfall in the region 11) prevalent wind direction, its velocity, m) wave action, if any.

Land Topography

The shoreland with reasonable stability and having a gentle slope towards the watershed is considered as most ideal for installation of the pen enclosure. Very steep gradient or closely spaced deep bed undulations pose difficulty in erection and operations. Hence such sites are unsuitable for pen installation.

Type of Soil

The soil should be productive and have sufficient bearing capacity to support the self-weight of the pen structure. The soil showing proneness to erosion or bed scouring is not suitable for the purpose.

Catchment Area and its Characteristics

Catchment area or catchment basin means the areas from which rainfall flows into the drainage outfall, lakes, beels etc. The boundary line of this basin is called the watershed. Surface run-off from the catchment area to the watershed depends on- 1) intensity and duration of rainfall 2) Land contours, land shape and area, 3) loss from evaporation, percolation and transpiration by vegetation etc. Floods from a large catchment area take longer time to rise than floods from a smaller catchment area.

Construction of Pen

Pens are constructed with the help of bamboo screens and nets. Split bamboo should not necessarily be shaped and rounded. They are soaked in water for two weeks and then dried for one week. During the soaking and drying period, bamboo poles are prepared and staked at the chosen site according to the desired size and shape of the fish pen. After stacking poles, bamboo splits are closely woven extending to a length of more or less five meters and made into a roll. After weaving, these are set by stretching them from one pole to another interturned or just set inside or outside close to the poles from bottom to top. They are tied every pole by rubber and one provided with sliced rubber around, limning one on top and one at the bottom. These splitted rubber prevent them from wear due to wave action.

Construction of pen made out of synthetic netting is easier than one made of bamboo screens. Fisherman can connect the nets into the fish pen after taking into account the desired height or depth of the pen site. After the net is constructed, the poles are staked in mud after making a provision for the front rope and the rope at the interval of 1.0 to 2.0m per stake and also the provision for float rope. In preparing the poles, all nodes are cleaned except one node with brunch protending one inch which is staked in the mud from 15-30 cm or more depending upon the depth of soft mud. With this node the foot rope is tied, and these together with the bottom net are staked in the mud. Boulders can be used as sinkers in the absence of lead sinkers. Bamboo tips of 1-11/2 m are also used to stake the bottom net with a foot rope firm into the mud to avoid escape of the fish stock. Construction of the nursery net may be done before or after the construction of the fish pen. They should have a free board of about 1 meter above the normal water level to prevent entry or exit of fishes

Figure 9.1: Pen Culture in Reservoir

by jumping and as a precaution against water level fluctuations. Metals and metal coated with HDPP screens are often used for pens which is highly durable.

Culture

Pen culture is extensively practices in Japan, Peru and Philippiness. Fish farmers in Laguna debay and Sansabo kekes stock milkfish fingerling in pens and grow them to marketable size. Prawns are also cultured in pens.

A pen culture technology has been developed by the Central Inland Fisheries Research Institute which is suitable for large scale adoption in the beels of West Bengal. Experiments have been conducted in the Akaipur beel of West Bengal to culture the giant freshwater prawn, *Macrobrachium rosenbergii* in pens.

The CIFRI technology is simple and inexpensive. The pen material is prepared from bamboo, which is locally available in plenty. Split bamboos are woven together with coir ropes. The split bamboo mats are erected in the beel and they are covered with close meshed nylon cloth. The bamboo screens are further reinforced with galvanized iron mesh for protection against crabs.

The pens were stocked with prawn juveniles of 75 to 80 mm size (4 g) at a stocking density of 12,000 prawns per ha. The prawns, harvested after a grow out period of 89 days, were found to grow up to 230mm length and 160 g in weight, the averages being 190mm and 86 g respectively.

Lime was applied in the pen water as a prophylactic measure against diseases. A locally manufactured feed was given to prawns to supplement the natural feed available in the pen. The artificial feed made of prawn meal, contained 23 per cent protein. Feeding was done during night @ 3-4 per cent body weight. The experiments are being continued. Preliminary results indicate that the feeding rate can be reduced further and a higher stocking density is possible. Stocking density, feeding rate and the species mix are being standardized.

In another experiment conducted by CIFRI, fish and prawn were cultured in pens. Two pens of 0.08 ha and 0.064 ha size were erected to conduct fish and prawn culture demonstrations in Mathura beel (flood plain wetland), North 24-Paraganas, West Bengal. The depth of water was 0.9 to 1.3 mm. Pens comprised 3 layers where bamboo splits formed the outer layers, iron mesh the middle layer and nylon mosquito net formed the inner layer.

The pens were stocked with fingerlings of catla (45 g), rohu (30 g), mrigal (27 g) and silver carp (42 g) at a stocking density of 10,000 nos/ha maintaining a ratio of C:M:R:SC:: 3:3:3:1. The fishes, which were fed on pelleted feed, were harvested after 130 days of operation, catla recorded an average growth of 400 g while rohu, mrigal and silver carp, 172,143 and 200 g respectively. Actual production of fish from a pen was estimated at 210 kg/0.08 ha/130 days.

In another pen M. rosenbergii weighing 5-8 g (82-100 mm) were stocked @ 25,000 nos/ha. The prawns were fed with commercial prawn feed, in addition to molluscan meat thrice in a week. Retrieval of prawn was only 56.39 per cent. The growth of the prawn on an average has been estimated at 41 g. A total of 37 kg of prawns was harvested at a yield rate of 578 kg/ha/130 days.

In experiments conducted by CIFRI in pens installed in Muzaffarpur lake (Bihar) C. catla, L. rohita and C. mrigala stocked in the ratio of 5:4:1 with an average weight 100 g achieved in six months average weight of 1 kg, recording a production of 4 tonnes/ha. The pen culture experiments at Kallai backwaters gave a production of 250 kg/ha in case of P. indicus. Similar experiments on P. monodon in Chilka lagoon gave a production of 100 kg/ha/3 months with 50 per cent survival. In Kakinada Bay, the blood clam, Anadora granosa was cultured in submerged pen at a density of 100 nos/m² and recording an yield of 385 kg/ha/5 months with a survival of 88 per cent. This yield is also in an impressive one. Pens were tried as an alternative for nursery ponds towards carp seed production. A bamboo enclosure of 250 m² fixed in the littoral area of Poongar swamp yielded advanced fry and fingerlings of C. mrigala and L. fimbriatus @ 1.27 million/ha in 90 days.

A production rate of 4 tonnes of fish in pens of 1 ha area in 6 months in an oxbow lake at Muzaffarpur in Bihar has already been achieved (Anon, 1983).

Yadava et al. (1983) recorded growth of common carp to be at 200 gm stocked at 10 nos m⁻² in a 20 m² pen in 5 months.

Rai and Singh (1986) experimented with two combinations in pens. In the first instance, the species ratio of rohu:catla:mrigal was

4:5:1 whereas in the second instance the respective ratio was 47:28:25. In both the experiments, they recorded a production of 4000 kg/ha indicating absence of any influence of species combination under identical conditions.

Bhowmik (1990) recorded a net fish production of 5545.5 kg/ ha from pen culture experiment of Indian major carps in a closed oxbow lake near Muzaffarpur city.

Adoption of pen nursery technology for the production of fingerlings using the peripheral areas of open water systems has immense application in the development of their fishery (Mathew, 1990). The massive extension of pen nursery technology in reservoirs would increase the present average yield rate of 14.5 kg/ha/yr to 45 kg/ha/yr and would result in an additional yield of 0.52 lakh tones from the presently available 17 lakh hectares reservoir area. Similarly, the large scale extension of pen nursery technology to other open water systems and riverine wetlands using their marginal areas and meeting the seed requiriements will help to achieve an additional yield of 3-4 lakh tones from the available 6 lakh ha of such waters. Further, it will also improve the socio-economic and nutritional rural populace and create additional employment opportunity to thousands.

Economics

Very little information is available on economics of pen culture in India. However, in this culture systems the capital costs vary with cage size and material used. Similarly operational costs are also variable as they are determined by species, site, method of culture and scale of operations. In India the technology is being standardised. Therefore accurate economics can't be worked out.

Advantages and Limitations

The advantages are numerous. It helps in economic and maximum use of available water resources thus reducing the pressure on land, facilities combination of different types of culture within one water body. It also has the merits of easy and economical control of predators and diseases, complete harvest of fish production and cutting down on the cost of preservation and transportation since they can be located in water ways and water areas near urban markets. The system helps in optimum utilization of artificial food for growth and maximizing its conversion rate to fish flesh. However, the limitations of the system are- difficult operation in rough surface water, high dependence on artificial feeding, need for rapid water renewal, increased risk of poaching and increased labour costs for handling, stocking, feeding and maintenance. In lakes and reservoirs, pens are restricted to the shallow zones near shores. This can interfere with the spawning of many species of fish and the shelter needed by the young fish. Because of the high concentration of fish, pens act as a magnet to fish eating birds, reptiles, and mammals. Cultured exotic fish frequently escape from pens, and this can introduce exotic species into the environment. Intensive pen farming has several effects on the water quality, including increase in levels of suspended solids and nutrients. There is also a decrease in dissolved oxygen in and around the enclosures (Beveridge, 1984).

References

Anon. 1983. Newsletter, Barrackpore, 6(5): 2

Beveridge, M. C. M. 1984. Cage and pen fish farming covering capacity models and environmental impact. FAO Fish Tech. Pap. 255-13pp.

Bhowmik, M. L. 1990. Pen culture-A means for higher fish yield from oxbow lakes. In: Contributions to the fisheries of Inland Open Water Systems in India (A. G. Jhingran, V. K. Unnithan and Amitabha Ghosh eds.), p. 85-88, Inland Fisheries Society of India.

CIFRI. 1996. Pen culture of prawns in the beels of West Bengal. *The Inland Fisheries News.* 1 (1): 1-2.

Mathew, Abraham. 1990. Pen nursery technology for the development of fisheries of oxbow lakes and reservoirs. In: Contributions to the fisheries of Inland open water systems in India, Part I, (A. G. Jhingran, V. K. Unnithan and Amitabha Ghosh Eds.), p. 89-92, Inland Fisheries Society of India.

Rai, S. P. and Singh, R. C. 1986. Happy in Pen. *Intensive Agriculture.*

Yadava, Y. S., Choudhury, M., Kolekar, V. and Singh, R. K. 1983. Pen farming-utilizing marginal areas of beels in Assam for carp culture. *Proc. Natl. Sem. Cage Pen Culture*: 55-58.

Chapter 10
Paddy-cum-Fish Culture

The culture of fish in paddy fields is of great significance in the economy of rural areas. It can provide a supply of cheap and wholesome protein food, besides affording an additional income. It is an old practice in several countries as Japan, Malaysia, Italy, China and India. There is a considerable scope for this practice in states of Manipur, Tripura, Assam, Bihar, West Bengal, Uttar Pradesh and Orissa where enough water is present in the paddy fields, which otherwise left unused during a certain period of the year. However, in recent years, interest in paddy-cum-fish culture has declined because of indiscriminate use of pesticides to protect high yielding varieties of paddy. These pesticides if used properly will in no way harm the fish population around the paddy fields.

Selection of Paddy Plot

Soil type is an important criterion for selection of paddy plot. Almost every type of soil like sandy loam to clay is suitable for paddy growing but clay soil preferred for the better paddy production and its high water retention capacity, ground water level, and drainage are the other features tobe considered before construction of plot. Therefore, the proposed plot having uniform contour and high water retention capacity is preferred.

Types of Paddy Plot to Integrate Fish Culture

The renovation of paddy plot depends according to the land contour and topography.

1. *Perimeter type*: Paddy growing area is in the middle with moderate elevation and ground sloping on all sides into perimeter trenches.
2. *Central pond type*: Paddy growing area is on the perimeter with sloping towards the middle.
3. *Lateral trench system*: Trenches are provided on one side or both sides of the moderately sloping field.

Preparation of Paddy Plot

The plot is prepared by sloughing the land at least twice during months of April –may before sowing the seedlings. Raw cattle dung @ 3000 to 5000 kg/ha is applied to paddy grown areas before ploughing. Direct sowing of the desired variety seedlings is wise, especially in the low-lying areas after first shower of monsoon. As the monsoon advances, the entire paddy plot and trenches became a single sheet of water. The kharif crop is harvested during months of November/December. The experiments were conducted by Central Inland Fisheries Institute (CIFRI), at Rahara Research Center and obtained 1.0 to 1.5 ton production of kharif paddy.

Fish Culture

Before sowing the kharif crop in the paddy plot, the unwanted grasses/plants should be completely removed from the paddy plot and its surrounding. To remove the trash and weed fishes, Mahua Oil Cake (MOC) at the rate of 250 ppm should be applied. The organic fertilizer, which was applied during the preparation of paddy field, should also be applied in waterways at the same rate along with the inorganic fertilizers. This will help in producing the planktons and benthos.

Fish Stocking

The fish species which could be cultured in rice fields must be capable of tolerating shallow water (<15 cm), high temperatures (up to 35°C), low dissolved oxygen and high turbidity. Fish seed is generally stocked during June to August. Indian Major Carp fingerlings *viz. Catla, Rohu* and *Mrigal* can be stocked at the rate of 5000 to 6000 nos/ha. Fishes like *Labeo bata, Puntius javanicus, Chanos chanos, Oreochromis mossambicus, Anabus testidineus, Mugil spp, Channa striatus, C. marulius* have also been found to respond well when cultured in the perimeter canal system. Culture of *Hypophthalmichthys molitrix* (Silver carp) and *Clarias batrachus* on the perimeter canal is also profitable

To increase the fertility of the trenches or canals during December to March, cowdung (5000 kg/ha), Ammonium sulphate (70 kg/ha) and Single Super Phosphate (50 kg/ha) may be applied in three equal doses. If available the domestic sewage water can also be applied for fish culture.

Supplementary Feeding

Supplementary food is given to enhance the fish growth rate. Groundnut oil cake and rice bran in the ration of 1:1 may be given to the fishes at the rate of 2.5 per cent of their body weight particularly after the harvest of khariff crop.

Fish Production

In this practice, the fishes are being cultured for a period of 5-6 months in the entire area and 4-5 months in the trenches. The fish production ranges from 500-700 kg/ha/year as obtained by CIFRI. Devraj and Natarajan (1973) stocked 50,900 fingerlings of different cultivable fishes in 11.9 hectare of paddy field and recorded production of 240-kg/ha area in 9 months period. According to Anonymus (1959 a) the *Cyprinus carpio* grow fast in paddy fields giving an average yield of 6.7 kg/acre, provided that the measures are being taken to prevent high rate of their mortality. Among the other carps the rohu and catla have better survival rate better rate but poor growth in comparison to the common carp (Anonymus, 1962). The average yield of tilapia is reported to about 31.3-kg/acre area of the field (Anonymous 1959 b)

Clarias batrachus is in great demand in the north -eastern parts of India, particularly in West Bengal, Assam, Orissa and Bihar for its high nutritional value. It contains higher percentage of protein and iron as compared to other edible fishes. It fat content is also very low and is therefore easily digestible so that it is very useful during convalescence. With most other it is a delicacy because of the characteristic aroma and softness of its flesh. Magur is available in plenty, if not abundantly during the post-harvesting period of 'Aman paddies' in the Sundarbans and other low lying areas of West Bengal. Magur is more profitable than that of any other cultivable species of fish. It is an ideal fish for culture in the paddy fields. Due to their omnivorous nature, they can consume a wide variety of insects as food. Moreover, they are hardy, easy to handle and economical to maintain. On the other hand, the high demand of magur, its fast growing nature, short culture period and use of paddy fields for rearing of fishes with selective and limited use of pesticides will commercially make the new system of integrated farming a success.

Besides this single tier system of integration of fish culture with other crops, there are other systems also where this integration may be 2 tier (fish-duck-horticulture) or 3 tier of 4 tier, depending upon the number of crops integrated. In China, a complex integrated farming is practiced in community ponds where live stock-piggery-duckery-poultry-floriculture-horticulture and agriculture are integrated with fish culture in large sized ponds to obtain maximum yield. But this type of complex integrated system requires efficient water and crop management skills; otherwise the whole system may be collapsed, because of high organic load in the ecosystem.

Though this system is not uncommon in China, but in India the simple tier integrated fish farming is even not popular and is yet tobe started on scientific commercial lines.

Control of Paddy through Biological and Chemical Means

With the introduction of high yielding varieties of paddy in different parts of India, the farmers have stared cultivating paddy in different seasons of the year from the same plot. But these high yielding varieties are prone to insect pests, such as rice hispa, stem borers, rice grasshoppers, brown plant hoppers etc. which are found to affect leaves, shoots and stem of paddy plants. The control of insect pests is very important as that of production management in paddy cum fish culture. The indiscriminate use of pesticides leads to the destruction of useful non-target species like birds, fishes etc. Some of the pesticides leaves toxic residues in crops, soil and fishes and poses a serious threat to human health.

Advantages in Paddy-cum-Fish Culture

1. Economical utilization of land.
2. Little extra labour.
3. Savings on labour cost towards weeding and supplemental feeding.
4. Enhanced rice yield.
5. Additional income and diversified harvest such as fish and rice from water, and onion and sweet potato through cultivation on bunds.

Problems in Fish Culture in Paddy Fields

Most common problem faced in paddy cum fish culture is the loss of fish during the period of their growth. The loss amounts to about 40 to 60 per cent for the young and 20 to30 per cent for the large fishes. It is due to birds like herons and others etc. living on the fish. Inadequate oxygen level, abrupt change in temperature and the poor depth of water also contribute significantly to the loss of fish.

Chapter 11
Fish-cum-Duck Farming

Introduction

Fish-cum-duck farming is an economically viable and technologically simple system and thus can be taken up by the rural farmers to improve their economic condition. The first scientific experiments on duck –cum- fish farming were made by Probst in Germany in 1934, but because of World War II the results remained unutilized. After the war, when there was a serious protein shortage in European countries, large scale experiments were conducted in Hungary and East Germany to determine optical husbandry methods for raising ducks on fish ponds. This practice of integration spread to other countries like India, Pakistan, Sri Lanka, Thailand, China, Bangladesh etc.

In the system the land area in the farm *viz.*, embankment, dykes etc. can be used for constructing duck pens, feeding ground for ducks as these areas harbor insects, worms, etc. Duck dropping can be used effectively for manuring pond and by rearing proportionate number of ducks; the necessity of manuring and supplementary feeding can be eliminated. Also ducks feed on aquatic insects, tadpoles, molluscs, snails, soft aquatic weed (lemna, azolla etc.) which are harmful to fish culture and thus help in keeping the environment clean. Again, in their search for food, ducks dabble the pond bottom, which helps in releasing obnoxious gases as well as nutrients from the pond bottom. Thus, in the system everything is effectively utilized for the production of valuable proteinecious crops. The expenditure spent on duck farming is largely offset by the sale proceeds of the duck flesh and eggs. In this system of farming about 3,500 to 4,000 kilogram of fish and 500 to 600 kilograms of duck meat and 18000 to18, 500 eggs can be obtained from a hectare pond area in a year.

Selection of Ducks

In selecting ducks two things should be borne in mind; sturdiness so as to withstand the weather conditions of the open pond in all seasons and the egg laying capacity so as to earn more income. Two types' Indian runner and Khaki cambel are found to be most suitable for this system of farming. The egg laying capacity of Indian runner is 150 eggs in a year and that of Khaki cambel is 180 to 200 eggs in a year and their growth rate is about 1 kilogram in a year.

Rearing Density

The object of duck raising in this system is to get duck easily to manure pond water under fish culture and thus the number of ducks should be adjusted so as to get adequate duck dropping to manure the pond. It has been found that 500 to 1500 ducks are sufficient to manure the pond of one hectare water area under composite fish farming with their dropping. Two to three week old ducklings should be introduced in the system after administering necessary prophylactic medicines to avoid mortality and much expenditure.

Feeding

Ducks are set free in the farm during the daytime and thus they are able to find their natural food such as insects, worms, tadpoles, molluscs, soft aquatic weeds, etc. from the pond water and the embankments. But this is not sufficient for their proper growth and hence artificial food may be given to supplement the natural food, by mixing any standard balanced poultry feed with good quantity rice bran in the ratio of 1:2 by weight. The feeding of this mixture may be done at the rate of 100 gm per each duck in a day in two times in equal halves-first one in the morning and another one at the evening either in the duck house or on the pond embankment so that the spilled feed can be easily drained into the pond. The feed should be mixed with enough water in the container so that the bill of the duck can be submerged at the time of tasking feed as the duck cannot feed properly without water.

Egg Laying

Ducks start laying eggs at the age of 24 weeks and continue to lay egg for two years. Generally the duck lays eggs at night. Each duck lays about 150 eggs, 180-200 eggs (Khaki cambel) in one year.

Harvesting

Harvesting should be done after two year rearing since after two years; the egg laying and fattening capacities are reduced considerably. It is rather more economical to dispose them off after one year rearing as they cost more than that age due to softness of their flesh. Egg collection can be done in the morning every day after lefting ducks out from the pen. It is expected generally to obtain 500 to 1500 kilograms of duck meat and 36,000 to 75,000 eggs in one-year rearing period.

Use of Duck Dropping as Manure

The duck droppings contain 25 per cent organic and 20 per cent inorganic substances, and are rich in carbon, nitrogen, phosphorus, potassium, calcium etc. Ducks are called as 'living manuring machines' and provide very good fertilizer to the fish pond and dabbling of ducks at the bottom in search of food releases nutrients from the soil, further enhancing production of pond. They feed on insects, larvae, tadpoles, molluscs, worms and aquatic weeds, which do not form food of stocked fish, and thus the two are highly compatible. Ducks are freely allowed to stay in the pond during the day so that their dropping fall automatically all over the pond. Dropping voided at night in the duck pens are washed off and the washing is drained into the pond every morning.

A duck voids on an average 125 to 150 gm droppings every 24 hours and thus 200 to 300 ducks would void on an average 10,000 to 15,000 kg of droppings during the year, which is sufficient to manure one-hectare water area under composite fish culture. The duck dropping has high manuring value.

Chapter 12

Role of Plankton in Fisheries

The term 'plankton' was coined by German biologists Victor Hensen in 1887. The term originated from Greek (literally that which wanders about) is a collective term for all small organisms which float on the water surface and drift at the mercy of water currents. Those of plant origin are called as the 'phytoplankton' and of animal origin are called as 'zooplankton'. Plankton can also be classified on the size as megaplankton (more than 8 cm), macroplasnkton (size vary from 1 mm to 1 cm), mesoplankton (0.5 to 1 mm), microplankton (0.06 to 0.5 mm), nannoplankton (0.005 to 0.06 mm) and ultraceston (0.0005 to 0.005 mm). The term picoplankton has been introduced recently for the exceedingly small organisms less than 0.0005 mm size. Plankton can also be classified on the basis of size. Net plankton are those retained by a tow net while nannoplankton will normally pass through.

Phytoplankton

Among biotic communities phytoplankton constitute the first stage in trophic level by virtue of their capacity to transduce environmental radiant energy into the biological energy through photosynthesis. Also referred to as primary productivity, the magnitude of photosynthetic energy fixation depends primarily on diversity and biomass of phytoplankton. The planktonic photosynthesis plays a key role in conditioning the microclimate (zone around an ecosystem) as it helps in regulating the atmospheric level of oxygen and carbon dioxide. Apart from primary production, phytoplanktons also play an important role as food for herbivorous animals. They also are biological indicators of water quality in pollution studies. To summarize, due to their environment in cycling of energy and matter in an ecosystem, evaluation of phytoplankton population in terms of their diversity, density, biomass, spatial and temporal distribution, periodicity and productivity and population turnover, is vital in management of an ecosystem. Fishes consume the phytoplankton, which is found abundantly in ponds, lakes, streams and reservoirs. Phytoplankton also gives green colour to the water. It is due to the presence of chlorophyll. Growth and multiplication of phytoplankton is mainly dependent on temperature, solar illumination and the availability of certain essential nutrients such as nitrates, silicates and phosphates. A chemical analysis of phytoplankton has revealed that they are nutritious as any vegetable. Phytoplankton used as food by fishes belongs to classes chlorophyceae, cyanophyceae, and Bacillariophyceae.

Class: Chlorophyceae

Due to the presence of chlorophyll these are known as 'green blue algae'. Many chlorophyceae members are useful as food of fishes. The chlorophyceae members useful as fish food are chlamydomonas, Volvox, Eudorina, Pandorina, Chlorella, Ulothrix, Oedogonium, Spirogyra, Microspora, Pediastrum, Cladophora, Clastridium, Scenedesmus, Cosmarium etc.

Class: Cyanophyceae or Myxophyceae

These are commonly known as 'blue green algae'. Under favorable conditions, blue green algae form extensive blooms, which can be toxic to other aquatic animals. The members of this class are Nostoc, Oscillatoria, Anabaena, Microcystis, Spirulina, Merismopedia, Arthrospira etc are consumed by the fishes. Myxophyceae dominate in many Indian reservoirs, the necessity of introducing fishes to utilize them as food was stressed by Natarajan (1975). The report of impressive growth of *Hypophthalmichthys molitrix* in Getalsud reservoir (Anon, 1977 b) and the high growth rate of the same fish in Kulgarhi reservoir (Natarajan, 1975) indicate that the fish can be successfully introduced to other myxophyceae-dominated lakes. Myxophyceae also dominated in the Yeldari reservoir of Maharashtra. The dominance of myxophyceae was also

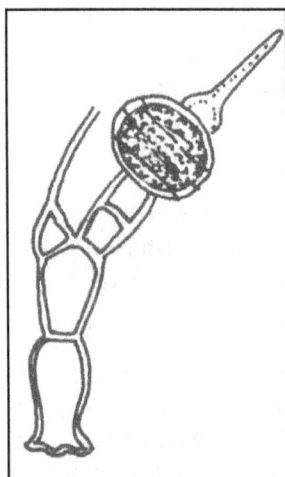
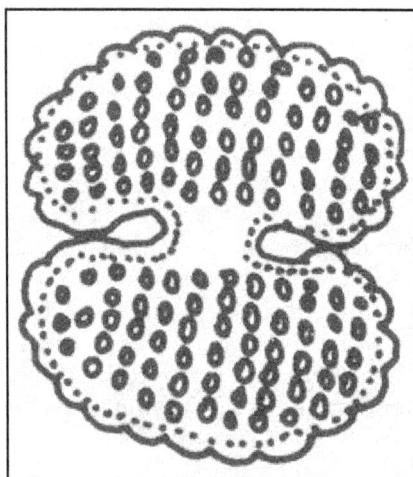

Oedogonium   Cosmarium

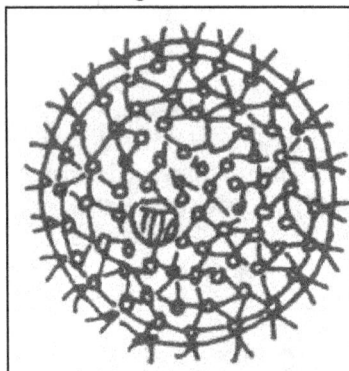
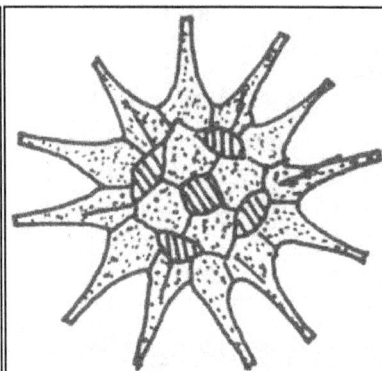

Volvox   Pediastrum

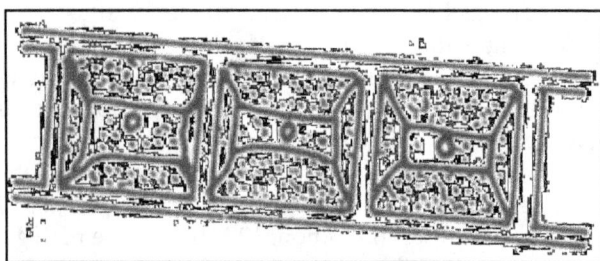

Ulothrix

Figure 12.1: Class: Chlorophyceae

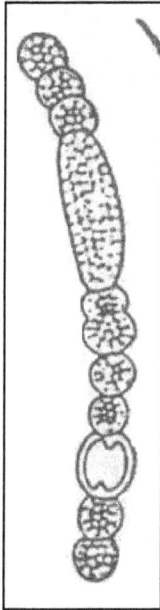
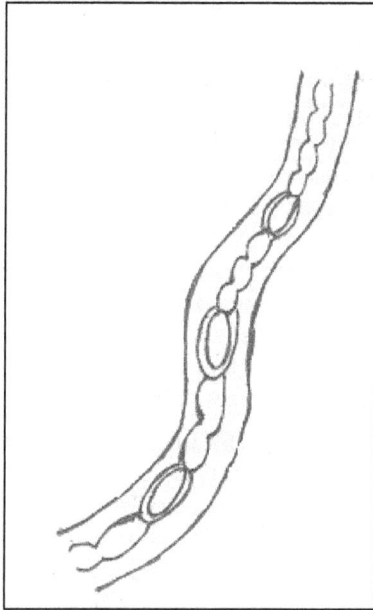

Anabaena   **Nostoc**

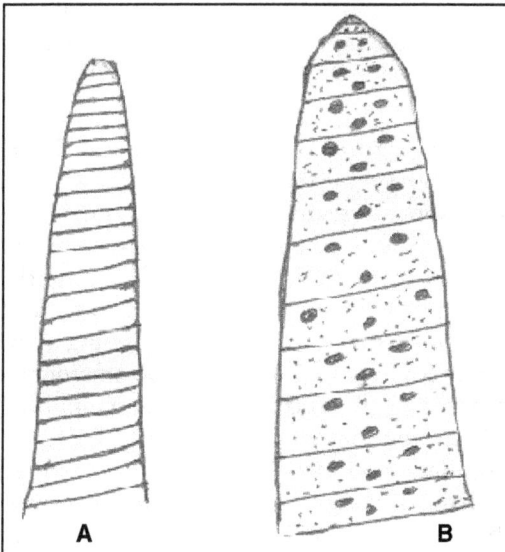

A   **B**

Microcystis

Figure 12.2: Class: Cyanophyceae or Myxophyceae

recorded by Rao (1982) in Manjira reservoir and Devi (1985) in Osmansagar and Himayathsagar, and found maximum domination in Mir-alam lake. Rao and Choubey (1990) reported that myxophyceae was third dominant group among phytoplankton in Gandhi sagar reservoir. Sugunan (1980) reported myxophyceae as the most dominant group followed by diatoms and green algae in Nagarjunasagar reservoir. He reported two peaks (October and April) of myxophyceae, which increased steadily till April and declined from May reaching minimum in August. Bacillariophyceae peaks were largely constituted by melosira. Navicula dominated the peaks in February and July. During the blooming of myxophyceae in summer, the diatoms were in a very few numbers and they picked up soon, after the bloom diminished (June-July).

Considering the nutritive potentials of spirulina it was incorporated in carp fry diets, replacing 10 per cent of the rice bran-groundnut oilcake mixture. The crude protein content of 23-25 per cent in the control diet had been increased to around 30 per cent by this addition. The mean specific growth rates of fish fed on the two diets were Catla 0.17,0.27; Rohu 0.19,0.63; Mrigal 0.54,0.73, Grass carp 0.02,0.40; and common carp 0.15,0.20. Large scale cultivation of spirulina is being considering, both as an input into fish/prawn culture systems as a feed ingredient and aneconomic enterprise for farm women for product export (Ayyappan *et al.*, 1991 d; Ayyappan, 1992).

According to Sugunan (1995) the ubiquitous blooms of microcystis in reservoirs in peninsular India are an example of a lacustrine biocoenose giving way to fluviatile ones in an impoundment. On the reservoir formation and the consequent transformation of lotic environment into the lentic system, saprophobes disappear from the scene giving room for the rapid multiplication of saprooxenes. Microcystis, finding a favbourable note with the new environment, bursts into blooms, out numbering all other forms into insignificance. In many reservoirs, orientation of lacustrine and fluviatile plankton can be clearly discerned from the composition of plankton in lotic, lentic and the cove sectors. The fluviatile lotic estor, although recording a lower plankton density, often shows better diversity and evenness indices, compared to the lentic and bay sectors, the still waters of which are characterized by higher concentration of dominance and low evenness (Sugunan.1991). Devi (1997) reported that the communities of

Cyanophyceae, Chlorophyceae, Bacillarioophyceae and Euglenophyceae constituting the phytoplankton bulk in Shathamraj and Ibrahimbagh reservoirs of Hyderabad. Among the phytoplankton community, the cyanophyceae was found to be rich and dominated in both the reservoirs.

Class: Bacillariophyceae

These are commonly called as 'diatoms'. They are unicellular organisms with different shapes and sizes. These may be yellow or golden brown or olive green in colour. The common diatoms consumed by fishes are Fragilaria, Nitzschia, Pleurosigma, Diatoma, Navicula, Cocconiers, Synedra, Tabellaria, Meridion, Amphipleura, Cyclotella, Melosira, Achnathes, Rhizosolenia etc.

Zooplankton

Zooplankton communities of freshwaters constitute an extremely diverse assemblage of organisms represented by nearly all the phyla of invertebrates. The most significant feature of zooplankton is its immense diversity over space and time. In an ecosystem 90 per cent of zooplankton species are herbivorous, remaining 10 per cent being carnivores. Since secondary production primarily depends on the biomass of herbivores, non-predatory zooplankters contribute significantly to the secondary productivity of an aquatic ecosystem. The diversity of zooplankton is usually studied by enumeration of different taxa or species in a representative sample collected by towing standard plankton net over an adequate distance. Zooplankton diversity is one of the most important ecological parameters in water quality assessment. The zooplankton study has been a fascinating subject for a long time. Water bodies rich in phytoplankton are also rich in zooplankton diversity and biomass. Among zooplankton, some of the organisms occasionally occur in appreciable numbers forming swarms. These swarms occur in freshwater ponds forming bands or streaks, or are arranged into areas of thick and thin concentration. Simulating cloud effect, they may give the water a strikingly different colour in the region of the swarm. The common zooplankton consumed by fishes belongs to protozoa, crustacea, and rotifera.

Protozoans

They were the first animals to evolve and thus have a special place in the evolutionary history of animals. These are found

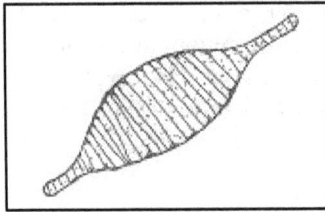

Nitzschia

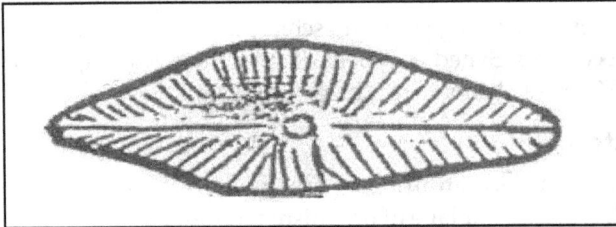

Navicula

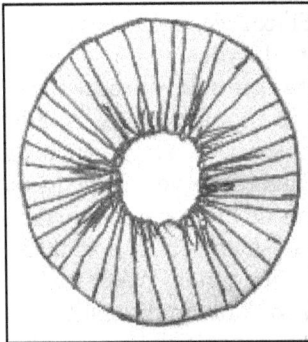

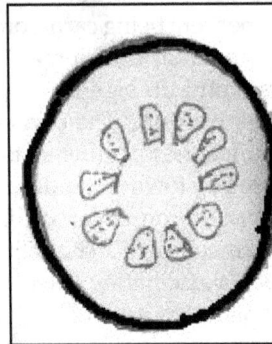

Melosira   **Cyclotella**   **Synedra**

Figure 12.3: Class: Bacillariophyceae

abundantly in fishponds and are useful as natural fish food. The different types of protozoans consumed by fishes are vorticella, Colopoda, Metropus, Euplotes, Oxytricha, Euglena, Amoeba, Ceratium, Chiomonas, Actinophyrys etc.

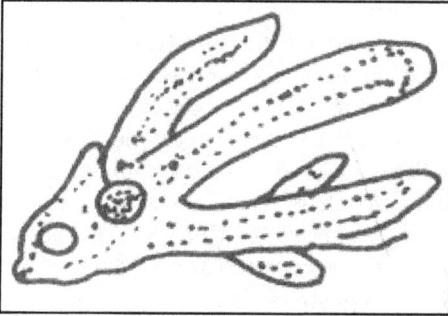

Amoeba

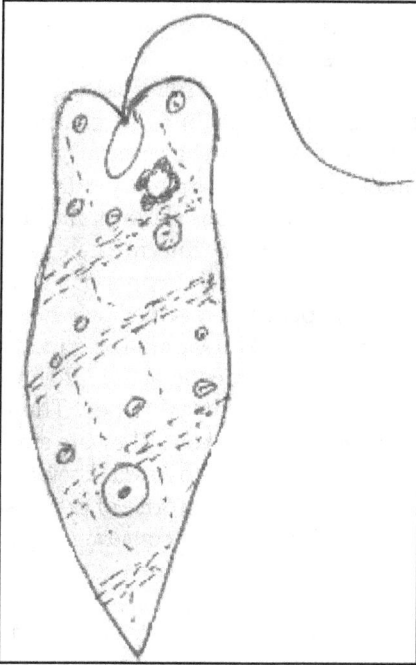

Euglena

Figure 12.4: Protozoans

Crusatceans

The aquatic animals with 19 pairs of appendages and branchial respiration are included in this class. The common crustaceans are the copepods, cladocerans, and the ostracods. Cladocera constitute an important group. The greater significance of cladocera in the aquatic food chain as food for both young and adult fish was emphasized much earlier. Generally the stomach content of young fish has 1 to 95 per cent cladocera by volume. Out of 11 families of cladocera,8 families have been reported from Indian waters, which represents about ont fourth of the world cladoceran fauna (Rao and Choubey, 1993).Nutrient rich cladoceran food is critical in the development of larval stages of culture fish species as during these developmental stages gastrointestinal tract as well as proteolytic enzyme systems are not well developed. Copepods are very small crustaceans. Their size varies from 0.3 to 2.5mm and body is cylindrical or pear shaped having long appendages. Among copepods, Cyclops, Mesocyclops, Diaptomus, Canthocamptus etc are used by fishes as their food. According to Devi (1997), free-living copepods are an essential link in food chain occupying the intermediate trophic level between bacteria, algae and protozoa on one hand and small and large plankton predators on the other. Though they are not as important as cladocerans in the diet of fish, they are well known as important intermediate host for helminth parasites in these reservoirs genre of copepods were recorded. A chemical analysis of copepods has revealed that they are proteinous as meat. The cladocearans like Daphnia, Ceriodaphnia, Moina, Eurycerus, Scapholebris, Daphniosoma, Polyphemus, Macrothrix, Alona etc are also used as food by fishes. The ostracods are commonly called as 'mussel shrimps'. They are small crustaceans and their size varies from 0.35 mm to 7 mm. Their body is enclosed in a bivalve carapace shell. They comprise of *Cypris* sp., *Stenocypris* sp. and *Cyprinotus* sp. which are also consumed by the fishes.

Rotifera

Dutrochet (1812) was the first to regard them as a separate biological group distinct from protozoa and called them rotifera. Over 2500 rotifer species belonging to 200 genera are known from various works in different countries and in India during the 20th century there are 40 researchers and 76 publications covering 300 and more species of rotifers belonging to 52 genera coming under 26 families (Dhanapathi, 000). Usually the rotifers like Rotaria, Lecane, Filinia, Trichocerca, Keratella, Brachionus, Polyarthra etc are useful as food organisms to fishes.

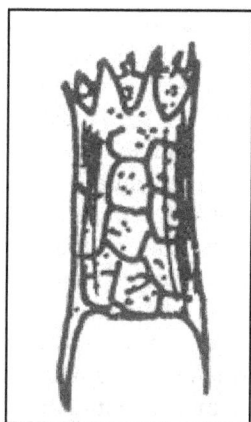

Brachionus

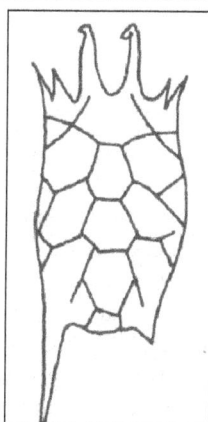

Keratella

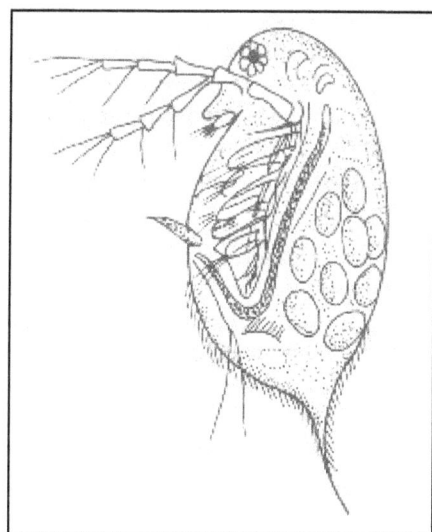

Daphnia

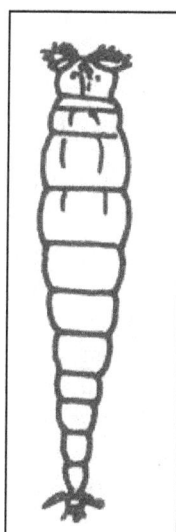

Rotaria

Figure 12.5: Rotifera

The density of plankton in a water body determines the stocking rate of fishes because they are the chief source of food of many economically important fishes. Plankton, due to its role in the ecosystem of the environment, is directly related to the fish catch potential of a reservoir. An insight into the distribution, composition and succession of plankton gives valuable clue for determining the fishing grounds, selection of suitable species for stocking and determining the level of utilization of the available food by the existing fish stock. Thus, the plankton study is essential for efficient fisheries management. Today, we have considerable knowledge on the very direct relationship between planktonology and fisheries. A stage has now considerable knowledge on the water quality and the nature of the plankton we can predict what the condition of the fishery in any area is going to be. Limnologists and fish biologists are now in a position to guide the fishermen to areas where good catches can be expected. Good fishing is no more a matter of chance.

Experiments have shown that 90 to 100 per cent fry stage of fish survive if the fry are stocked during swarms of zooplankton and abundant food is available. Fry prefers zooplanktons as they are easily digested. Phytoplankton is not utilized as food as is difficult to digest. The nutritional value of phytoplankton is also less than zooplankton. Application of heavy dose of cow dung in ponds usually results in early production of zooplankton swarms, and the stocking of the fry has to be adjusted with the time of its maximum production.

Culture of fish food organisms (diatoms, rotifers, cladocerans) and provision of periphytic substrates, are the aspects of trophic enrichment being considered. Periphytic substrates, like palm leaves and straw mats, were evaluated for their effects on biotic enrichment and fish production levels in carp culture ponds. In monoculture of Rohu, substrates provided to the extent of 7 per cent of pond area showed settlements of Volvox, Pediastrum, Scenedesmus, Ceratium, Diatoms, Brachionus, Keratella, Asplanchna, Bosmina and Ostracods to the extent of 763-1131/cm² that was further reflected in fish production levels, being 10.05-11.71 per cent higher than the control ponds (Ayyappan, 1994).

Information is available on the temporal and spatial distribution of plankton from many water bodies in the country. Sugunan (1980) studied the plankton of Nagarjunasagar reservoir of Andhra Pradesh and reported the seasonal abundance of planktons. He observed the gut content of fishes from the same reservoir, which indicated a poor utilization of major plankters as food by commercial fishes. Plankton constitutes a vital link in the aquatic food chains. While phytoplankton plays a phenomenal role in bio-synthesis of organic material. Zooplankton, an important component of secondary production, provides a link between the producers and secondary consumers. A scientific appraisal of the spatio-temporal variation of plankton community provides information necessary for a proper understanding of growth and abundance of fish yields and also whether there is any scope for introduction of addition species of commercial value in order to utilize the vacant food niches, if any. The most significant feature of zooplankton is its immense diversity over space and time. Thus, similar aquatic systems may have dissimilar assemblage of organisms varying in species composition and biomass. Further inspite of convergent similarities zooplankton species have different types of life histories influenced by seasonal variations of abiotic factors, feeding ecology and predation pressure. Zooplankton diversity is one of the most important ecological parameters in water quality assessment. Various indices like richness, diversity and evenness index can be calculated when data on taxonomy of different zooplankter is available. It is found that zooplankton has an inverse relationship with water temperature. The possibility that the zooplanktons avoid bright sunlight and high temperature of surface water has been put forth by Russel (1927) and Clarke (1933). But temperature alone may not control vertical movements of zooplanktons. Hardy (1936) stated that zooplankton migrate up to the surface water to feed upon phytoplankton during night but when there is abundant phytoplankton, zooplankton soon migrate downward again to a great depth, owing to the effect of 'animal exclusions' on phytoplankton.

References

Anon. 1977b. Central Inland Fisheries Research Institute Newsletter, 1(3): 4.

Ayyappan, S. 1992. Potentials of spirulina as a feed supplement for carp fry. In: Spirulina ETTA Nat. Symp., Seshadri, C. V. and Jeeji Bai, N. (Eds.), MCRC, Madras, pp171-172.

Ayyappan, S. 1994. Micrbiological technology for aquaculture. Proc. MICON-Ayyappan, S., Pandey, B. K., Sekar, S., Saha, D. and Tripathi, S. D. 1991d. Potenitials of spirulina as a feed supplement for carp fry. Proc. Nat. Symp. Freshwat. Aqua., pp 86-88.

Clarke, C. L. 1933. Diurnal migration of plankton in the gulf of manio and its correlation with the change in submarine irradiation. Biol. Bull. 65: 404-436.

Devi, M. J. 1985. Ecological studies of the limnoplankton of three freshwater bodies of Hyderabad. Ph.D. Thesis. Osmania University, Hyderabad.

Natarajan, A. V. 1975. Fish farming in man-made lakes. *Indian Fmg.* 25(6): 24-33.

Devi, Sarla B. 1997. Present status, potentialities, management and economics of fisheries of two reservoirs of Hyderabad. Ph. D. thesis, Osmania University, Hyderabad.

Dhanapathi, M. V. S. S. S. 2000. Taxonomic notes on the rotifers from India (from 1889-2000). Indian Association of Aquatic Biologists, Hyderabad, Publ. No. 10.

Hardy, A. C. 1936. Plankton ecology and the hypothesis of animal exclusion. Proc. Linn. Soc. Lond. Session, 148, 1935-36t. 27: 64-70.

Rao, I. S. 1982. Ecology of the Majira reservoir. Ph. D. thesis, Osmania University, Hyderabad.

Rao, K. S. and Choubey, Usha. 1990. Studies on phytoplankton dynamics and productivity fluctuations in Gandhisagar reservoir, p. 106-112. In: Jhingran, Arun G. and Unnithan, V. K. (eds.). Reservoir Fisheries in India. Proc. of the Nat. Workshop on Reservoir Fisheries, 3-4 January 1990. Spl. Publ. 13, Asian Fisheries Society, Indian Branch.

Russel, P. S. 1927. The vertical distribution of animals caught in the ring-trawl in day time in the Plymouth area. *J. Mar. Biol. Ass. U. K.*, 14: 557-608.

Sugunan, V. V. 1980. Seasonal fluctuations of plankton of Nagarjuna reservoir, A. P., India. *J. Fish. Soc. India* 12(1): 79-91.

Sugunan, V. V. 1991. Changes in the phytoplankton species diversity indices due to artificial impoundment in river Krishna at Nagarjunasagar. *J. Inland Fish. Soc. India,* 23(1): 64-74.

Sugunan, V. V. 1995. Reservoir Fisheries of India, F. A. O. Fisheries Technical Report 345, Daya Publishing House, Delhi.

Chapter 13

Food and Feeding Habits of Fishes

The food plays one of the most vital roles in the life history of fishes by way of controlling their growth, fecundity and migration. Variation in seasonal and diurnal availability of the preferred food organisms of various fish species in any region may govern the horizontal and vertical movements of the fish stocks. A study of the food and feeding habits of fish is very important in any fisheries research programme. Assessment of the food and feeding habits of fish help us to determine its niche in the ecosystem and its preferred food items. It tells us how much the food spectrum of a fish overlaps with that of the co-existing fishes. The analysis of food components in the gut of a species from different habitats provides the information on how much the species is selective in the choice of food and how flexible it is in feeding on different food items.

Fishes are also known to change the food habits as they grow, accompanied by correlative changes in the digestive system. Another remarkable feature regarding the fish feeding is its seasonal variation in selection of food in the biotope. Due to this, different dietary pattern of the same fish from various habitat have been reported. Some fishes feed on plants (herbivorous), others feed on animals (Carnivorous), while a large number of fishes derive their requirements from animals as well as plants (Omnivorous). A few fishes depend for their nutrition entirely on zooplankton and phytoplankton. These fishes are known as plankton feeders. The plant material consumed by fishes included unicellular algae, filamentous algae, and portions of higher aquatic plants. The animals consumed by fishes are the mussels, insects, larvae, smaller fishes, worms, crustaceans, frogs and tadpoles.

The gill rakers are important structures, which enable the fishes to strain food materials (plankton) from ventilating water and divert the same to the oesophagus for consumption. The rakers are important structures used for straining food and other materials from water ventilating through the gill sieve (Munshi *et al.*, 1984, Ojha *et al.*, 1987). They are designed to to suit the food and feeding habits of the fishes. The taste buds on the free margins of rakers are associated with detecting the chemical structure of food and water ventilating through the gill sieve. Chemoreceptor function of the taste buds of the rakers is also reported in the mullets (Hossler *et al.*, 1979, Ojha *et al.*, 1987).

The natural food of fishes can be divided into following types:

1. *Main/Basic food*: It is the natural food consumed by fish under favorable conditions.

2. *Occasional/Secondary food*: It is eaten by the fish in small quantities when available.

3. *Incidental food*: This type of food rarely enters the gut along with other items.

4. *Emergency/Obligatory food*: It is ingested in the absence of basic food and on which the fish is able to survive.

The food of young ones is generally different from that of the adult. Young ones with small and short intestine prefer zooplankton, and are able to digest rotifers, cladocerans another microscopic animals easily. The phytoplankton and algae are not easily digested.

Freshwater fishes are divided into three groups according to the niche they occupy in the water. These are surface feeders, mid or column feeders and bottom feeders. The surface feeders are *Catla catla, Puntius ticto, Barilius, Chanda nama, C. ranga,* and *Hilsa ilisha.* Some fishes are neither true surface feeders nor true bottom feeders and mostly depend on the organisms of the midwater. Such fishes are called as column feeders and include *Labeo rohita, Puntius sophore, Tor tor, Wallago attu, Mystus cavassius,* and *M. vittatus.* The bottom feeders depend mainly on bottom organisms. Fishes belonging to this group are *Labeo calbasu, Labeo bata, L. gonius, Cirrhina reba, C. mrigala, Puntius sarana, Amblypharyngodon mola, Clarias batrachus, Channa marulius, C. striatus, Heteropneustes fossilis* etc. Of these the surface feeders are either omnivorous or carnivorous, while the mid and bottom feeders may be herbivorous, omnivorous or carnivorous.

Seasonal variations and fluctuations have been observed in the available food for freshwater fishes, resulting in seasonal growth. Herbivorous fishes get an abundant food supply during April, May and June, and these fishes have their stomach full of algal food, but during winter season there is scarcity of food. Among the carnivorous fishes April, June, September and October are the good months for feeding, while July. August and December constitute the slack period. Fluctuation in the availability of food and feeding intensity results in seasonal growth of fishes. The omnivorous fishes do not show a marked seasonal variation in feeding activity due to the availability of one kind of food or the other throughout the year. Some freshwater fishes and their food and feeding habit are discussed below:

Heteropneustes fossilis

Insect larvae, insects and insect eggs formed the major food items of this species. In specimens collected from Aligarh, the gut contents consisted mainly of crustaceans followed by molluscs, insect larvae, small fishes, plant materials etc. and in the species from Punjab the major food item was crustaceans followed by oligochaetes, fish, insects and insect larvae.

Channa punctatus

Insects, and insect larvae formed the main food in Hussainsagar specimens (Siva Reddy, 1985), while in fishes from Guntur (Reddy, 1980) and Aligarh (Qayyum and Qasim 1964) fishes formed the main food item. Since *C. punctatus* is a carnivorous fish, insects and insect larvae naturally dominated among the gut contents (Baburao, 1990). The oligochaete annelids formed the other important component of the food of Hussainsagar murrels.

Channa punctatus is mainly a carnivorous fish. Its food consists mainly of crustaceans, insects, mollucs, fishes, plant and semi-digested materials. The juvenile and adult *C. punctatus* are surface feeders. Monthly variations in the percentage composition of the food items in both stages of the fish were recorded. The fish changed its food and feeding habit seasonally. The feeding intensity was very poor in mature fishes during the spawning period. The juvenile fishes fed actively throughout the year. The ratio of the total length and alimentary canal length of juvenile and adult stages is 1:0.36 and 1:0.28 respectively (Bhuiyan *et al.*, 2006).

Notopterus notopterus

A study of the feeding habits of fish from Hemavathy reservoir indicates that the major stomach contents of this fish includes insects (42 per cent), semi-digested organic matter (24 per cent), fish remains (16 per cent), prawns (7 per cent), miscellaneous items (7 per cent) and 4 per cent plant tissues (Sugunan, 1995).

Wallago attu

This fish is extremely voracious carnivore, feeds on fish. Insects, crustaceans and algae are sometimes encountered in the gut. Fingerlings consume insects, other fish fry and fingerlings. Major stomach contents include fish (82 per cent), insects (6.5 per cent), miscellaneous item (5.4 per cent) and 3 per cent decayed matter (Sugunan, 1995).

Chanda nama

Copepods, mainly naupli, Cyclops, Diaptomus (65 per cent), insects, mainly dipteran larvae, chironomids, mayfly nymphs (17.5 per cent); semi digested organic matter (15 per cent), fish remains (2.5 per cent) comprise the main food items in species recorded from Nagajunasagar reservoir (Anon, 1982).

Puntius pulchellus

The gut content analysis of *P. pulchellus* revealed that it is a herbivorous fish having preference for soft vegetation like *Vallsneria* and *Chara*. In India *P. pulchellus* is the only indigenous carp species, which feeds upon aquatic vegetation as well as submerged grasses. In absence of grass leaves, this fish consumes grass roots and if this too becomes scare, it feeds upon decaying vegetable matter and bottom debris. Although adults fed upon small sized gastropods, the fingerlings feed upon filamentous algae along with insects

(Routray *et al.*, 2001). One of the most remarkable features of the species is that, although vegetative, its gut length is hardly 3.0 to 3.7 times the total length of an adult (David and Rahman, 1975).

Puntius melanampyx

This fish could be categorised as bottom feeders(Anna Mercy *et al.*, 2002). It is omnivorous fish which consumes a wide range of food materials like detritus, filamentous algae, plant matter, crustaceans, and insects. The distinct preference to benthic flora and fauna is probably a reflection of the behavior of the species which spend most of the time in benthic zone, as the material in the digestive tract faithfully reflects relative environmental densities of food items falling within the ingestible size range (Anna Mercy *et al.*, 2002).

Puntius carnaticus

Young fry feed on phytoplankton (55 per cent), zooplankton (10 per cent) and aquatic insects (35 per cent). Adults feed on higher plants, filamentous algae, crustaceans, insects and fish remnants etc (Alikunhi, 1957).

Puntius dubius

It is bottom feeder, feeds on algae along with mud, gastropods and insects.

Puntius kolus

Epithemia, Amphora, Surirella, Synedra, Tabellaria, Diatema, Eudorina, Spirogyra, Diploneis and *Mastogloea* formed the major food item in the species collected from Bhatghar reservoir of Maharsashtra (CIFRI, 1997). According to Anon (1982) diet of the species comprise molluscs, mainly gastropods, bivalves (60 per cent); chlorophyceae, mainly *Spirogyra, Pediastrum* (3 per cent), diatoms, mainly *Fragilaria, Melosira, Navicula* (5 per cent), organic detritus (30 per cent), zooplankton, mainly copepods (2 per cent) comprise the main food items of species (Anon, 1982).

Osteobrama cotio

Organic detritus (70 per cent), zooplankton, mainly *Diaptomus, Cyclops, Nauplii, Chydorus, Keratella, Lecane, Daphnia* (20 per cent), and phytoplankton, mainly *Navicula, Fragilaria, Cymbella, Spirogyra* and *Pediastrum* (10 per cent) formed the major food items in species collected from Nagarjunasagar reservoir (Anon, 1982).

Cirrhinus mrigala

The blue green algae, dominated by Microcystis, constituted the major item of the fish, closely followed by the detritus. The detritus along with decayed organic matter constituted almost half of the gut contents (41.8 to 49.4 per cent) indicating the bottom feeding habit of the species (Selvaraj *et al*, 1997).

Cirrhinus reba

Appears to subsist largely on planktonic algae which form about 60 per cent of the food. Considerable quantities of mud, sand and debris are met with in the gut contents. The youngones feeds voraciously on zooplankton and grow faster than even youngone of catla. The limited quantity of decaying leaves and debris found in the gut probably indicates its more pronounced plankton feeding habits. However, as mud and sand are also almost invariably found in the gut, it may be inferred that the fish occasionally feed in the marginal shallows also.

Labeo fimbritus

The fish proved to be an absolute bottom feeder mainly subsisting on detritus (42.9 to 59 per cent) followed by the decayed organic matter, sand and silt (Selvaraj *et al.*, 1997). Adults feed on filamentous algae, together with protozoa, rotifers and copepods (Das, 2011).

Labeo calbasu

Only blue green algae (36.2 to 55 per cent), detritus (30 to 42 per cent) and decayed organic matter.

Labeo rohita

Detritus formed the bulk (56.1 to 58.8 per cent) of the gut contents. Blue green algae dominated by Microcystis14.6 per cent) and the decayed organic matter (10.2 per cent) occupied the second and third potion (Selvaraj *et al.*, 1997).

Labeo kontius

It is a native of Cauvery and Coleroom rivers of South India. Being a vegetarian, it feeds on algal filaments, diatoms and pieces of higher plants, copepods, rotifers, although causally feeding on insects and worms also occasionally.

Catla catla

Pinnularia, diatoms and detritus formed the major food items of this species (CIFRI, 1997). The work on this species from Aliyar reservoir has recorded blue green algae as the dominant item (43.8 to 48.1 per cent) followed by the detritus (21.3 to 24.7 per cent). The zooplanktons were restricted to 121.7 to 24.5 per cent consisting of copepods and rotifers (Selvaraj *et al.,* 1997).

Cyprinius carpio

The algal components dominated with 36.6 to 62.6 per cent among the gut contents. Detritus also had significant contribution (9.2 to 31.6 per cent) indicating its bottom browsing habit. Zooplankton and diatoms also formed the components of its diet to some extent (Selvaraj *et al.,* 1997). A slight difference in food of different size groups of the fish was noted by Sakhare (2010). The preferred items for small size groups were rotifers (62 per cent), copepods (26 per cent), diatoms (6 per cent) and blue green algae (6 per cent), while in the case of food of larger fishes, blue green algae (47 per cent), diatoms (28 per cent), detritus (14 per cent), rotifers (6 per cent) and decayed organic matter (5 per cent) were common. A certain variability in the feeding intensity was observed. In the female common carps there was a gradual drop in feeding intensity during the maturation phase (January-February and July-September). In the case of the male common carps the feeding intensity did not seem to be affected during maturation phase. According to Dube (2003) common carp is a polyphagus and omnivorous in habit. Initially feeds on plankton and crustaceans. The fingerlings gradually shift to benthic organisms such as chironomids, tubificids, insect larvae and decayed vegetative matter. The adults are omnivorous, preferring insects, crustaceans and decayed organic matters. In search of benthic food it burrows bottom and embankments of ponds thus making them weak and water turbid. Its feeding habit (browsing nature) could undermine river/reservoir banks leading to the collapse of banks and uprooting vegetation bringing changes to river flows/ courses. The foraging behavior of common carp resulted in vegetation removal both by direct consumption and by uprooting due to its proclivity to dig through substrate in search of food. The latter activity also resulted in increased water turbidity rendering the conditions more conducive for its propagation (Lakra and Singh, 2007).

Saikia and Das (2008) recorded a total of 60 food items of which 22 belonged to chlorophycea, 12 to the cyanobacteria, 10 to the bacillarophycea and 16 to several zooplankton taxa. According to Piska (1999) common carp is an omnivorous bottom feeding fish. Adult fish feed on bottom dwelling aquatic animals *viz.* insect larvae, worms, molluscs and decayed vegetable matters. It feed on epiphytic plankton also. Young feed on protozoan and small crustaceans.

Hypophthalmichthys molitrix

From early fry to late fry stage it feeds on zooplankton as main food with phytoplankton as occasional food but later on phytoplankton becomes the major food with zooplankton as occasional food. Phytoplanktivorous surface feeder from fingerling to adult stage, its main food is euglenoids, microcystes, nostoc, diatoms, desmids and filamentous green algae. Throughout life it feeds as occasional food on zooplankton

Ctenopharyngodon idella

It feeds on soft and hard aquatic weeds. It also accepts terrestrial grass growing on the bundhs. Though it has established the name as grass carp, in fact it is an omnivore. Fry feeds on organisms like cyclops, diaptomus, daphnia etc. However in the later periods, its food habit changes towards aquatic weeds. It feeding rate is completely different and extremely higher than other carps. It needs a minimum food of 25 per cent of its total body weight daily. Its maximum daily feeding ration has been as high as eight times of its total weight. Its habit of feeding aquatic weeds is beneficial in biologically controlling the aquatic weeds and it also serves as a 'living green manuring machine' besides its own growth. The feeding intensity during post spawning months is high and the adults consume Hydrilla, Najas, Vallisneria, Utricularia and soft leaves of Eichornia.

Aristichthys nobilis

It is a filter feeder. Only bigger live food organisms are retained for its consumption during the process of filteration. Generally it feeds on both phytoplankton and zooplankton, resembling the food habits of rohu.

Tilapia mossambica

It is a omnivore. The fry feeds exclusively on diatoms and other unicellular planktonic and epiphytic algae. The adults subsist mainly on vegetable food chiefly of chlorophyceae, myxophyceae and bacillariophyceae. When vegetable food are scare, worms, insects, crustaceans, fish larvae and detritus are eaten. Diatoms and chlorophyceae form the major items of food besides weeds in the case of young tilapia.7-10 mm fry feeds almost entirely on zooplankton. However, some algal items may occasionally be found in the stomach.11-60 mm fish feeds almost equally on zooplankton and phytoplankton. Above 60 mm, their main food is phytoplankton, filamentous algae and rarely leaves of higher aquatic plants (Sundararaj and Srikrishnadhas, 2000).

Osphronemus goramy

Gourami subsists largely on aquatic plants like water lilies, lotus, submerged weeds, surface creepers, marginal grasses, algae, insects, worms and prawns.

Anabus testudineus

It is omnivorous. Fingerlings feeds on animalcules, insect larvae, and water fleas, while adult prefer insects, water fleas, vegetable debris, fish etc.

Clarias batrachus

It is basically carnivorous (mainly insectivores) in habit. They prefer to eat decayed protein food and are considered as scavengers in fishpond. In cultural operation, they adopt themselves excellently to supplementary feeding with fishmeal, oil cake, rice bran, earthworms, insect larvae, silkworm pupae etc. The adult subsists on insect larvae, shrimps, worms, small fish and organic debris found in pond bottom. Young fry feed on protozoans, small crustaceans, rotifers, and phytoplankton.

Clarias gariepinus

It is carnivorous in feeding habit and fish is the most important food item. It contributed 81.7 per cent of the food items of the juveniles and 86.8 per cent of the adults by volume (Elias, 2000). *Orochromis niloticus* was the most utilized prey of *C. gariepinus*. *O.niloticus* accounted for 71 per cent of the food eaten by juvenile fish (16.3 to 35 cm TL) and 77.5 per cent of the food of adults by volume. Other food items found in the stomachs of *C. gariepinus* include insects, fish eggs, gastropods, pieces of macrophytes, detritus and zooplankton.

Mystus seenghala

It is a bottom and column feeder, predominantly a carnivore right from the advanced fry stage to adult. Fish is the main food item though it consumes good quantity of insects depending upon the availability in the environment during different seasons.

Mystus aor

It is a zooplankton feeder at an early stage but the feeding habit changes to animal organisms (fishes, insects, molluscs etc) in the adult stage. Juveniles are mostly insectivorous and marginal feeders.

Garra lamta

Food scrapers of *G. lamta* in the form of curved teeth like structures are important findings occasioned by scanning electron micrographs scraped algae and periphyton from submerged substratum is the basis of location of Garra. The gut analysis revealed that *G. lamta* feeds mainly on algae belonging to the families chlorophyceae, Xanthophyceae and Bacillariophyceae (diatoms). Leaves of some hydrophytes, decayed organic matters and sand granules were also obtained from the gut and designated as miscellaneous.

Nemacheilus rupicola

This fish has a large terminal mouth. Because of the carnivorous feeding habit the elaborate feeding apparatus in the form of food scrapers are absent. The maxillary and pharyngeal teeth are present. The food items of *N. rupicola* are identified into eight groups. Chlorophyceae, Xanthophyceae, Bacillariophyceae (diatoms), cladocerans, copepods, ostracods, insects and miscellaneous are the eight groups of food items obtained from the gut of *N. rupicola*.

Botia birdi

Kant and Vohra (1993) studied the gut contents to find out the quantitative and qualitative composition of the different algal components. The major dietary of this fish is animal food and the fish is a carnivore. Insects and their larvae constituting about 62 per cent the gut contents from the major part of the food of this fish followed by 15 per cent daphnids, 4 per cent crustaceans and 2 per cent rotifers. In all 83 per cent diet of the fish is of animal origin. Of the rest 17 per cent of the digested green matter contributes 4 per cent. Only 13 per cent algae is found in the stomach of the fish, of which cyanophyceae is the major constituent (5 per cent), followed by 4 per cent each of desmids and diatoms and green filamentous algae. The algae in this fish is not digested giving an indication that probably this does not form the basic food of the fish or the fish does not relish it much and perhaps these algae enter inadvertently into the suctorial mouth of the fish. Even though algae do not form a direct food of the fish, but it has been proved that the insects and daphnids along with rotifers and crustaceans do feed on the algae thereby forming an indirect food for the fish.

Anguilla bengalensis

It is a carnivore and mainly feeds on other fishes and invertebrates. Being nocturnal, it feeds only during night. Cannibalism is observed, when enough feed is not available in surrounding.

Chanos chanos

The young larvae feed on algae belonging to bacillariophyceae, myxophyceae and chlorophyceae. Fry and fingerlings feed upon diatoms, algae, lamellibranchs, fish eggs etc. It is primarily a phytoplankton feeder.

Mugil cephalus

This is filter feeder, feeding on organisms low in food chain.

Etroplus suratensis

Advanced fry take in plenty of aquatic insect larvae but when they are about 19 mm long, there is a shift in feeding habit and start feeding on filamentous algae and vegetable matter. Adults are herbivorus feeding on myxophyceae, chlorophyceae and decaying organic matter.

Hilsa ilisha

Hilsa is mainly a plankton feeder. Hilsa fry (20-40mm) mainly feed on diatoms, copepods, daphnia and ostracods whereas the younger hilsa (up to 100mm) feeds on smaller crustaceans, insects and polyzoa. Hilsa feed at the bottom. Many workers concluded that the food of young hilsa is dominantly plankton feeder and the species feed at all depths of water either in the freshwater zone or in the tidal zone of the estuary. Studies on the morphology, anatomy and histology of the alimentary canal and its associated structure like gill-rakers and pharyngeal organ revealed that the young hilsa mainly subsists on zooplankton while adults are microphagus.

Lates calcarifer

It is a carnivorous and predatory fish. It ascends to estuaries and backwaters in pursuit of food and feeds on small fishes, crustaceans, worms and snails.

References

Alikunhi, K. H. 1957. Fish culture in India. Indian Council of Agricultural Research, New Delhi.

Anna Mercy, T. V., Raju Thomas, K. and Jacob Eapen. 2002. Food and feeding habits of *Puntius melanampyx* (*Day*) -An endemic ornamental fish of the Western Ghats. In: Riverine and Reservoir Fisheries of India (Boopendranath, M. R., Meenakumari, B., Joseph, J., Sankar, T. V., Pravin, P. and Edwin, L., Eds.), p. 172-175, Society of Fisheries Technologists (India), Cochin.

Anon, 1982. Final Report of All India coordinated Project on Ecology and Fisheries of Freshwater Reservoirs, Nagarjunasagar. Research Information Series, 3 March 1983. Central Inland Capture Fisheries Research Institute, Barrackpore, West Bengal, India, pp 148.

Babu Rao, M. 1990. Effect of eutrophication on the biology of fishes in reservoirs-A case study at Hussainsagar. p. 48-52 In: Jhingran, Arun G. and V. K. Unnithan (eds.). Reservoir Fisheries in India. Proceeding of the National Workshop on Reservoir Fisheries, 3-4 January, 1990. Special Publication 3, Asian Fisheries Society, Indian Branch, Mangalore, India.

Bhuiyan Abdus Salam, Shamima Afroz and Tanjeena Zaman. 2006. Food and feeding habit of the juvenile and adult snakehead, *Channa punctatus* (Bloch). *J. Life Earth Sci.* 1(2): 53-54.

CIFRI. 1997. Ecology and Fisheries of Bhatghar Reservoir. Central Inland Fisheries Research Institute, Barrackpore (West Bengal). Special Publication No. 73.

Das, P. 2011. Diversification of freshwater aquaculture species: Potential of endemic fishes. *Fishing Chimes.* 31(1): 21-22.

David, A. and Rahman, M. F. 1975. Studies on some aspects of feeding and breeding of *Puntius pulchellus* (Day) and its utility in culturable waters. *J. of Inland Fish. Soc. of India.* Vol. VII Dec. pp 787-88.

Dube, Kiran. 2003. Biology, reproductive biology and embryonic development of carps. In: Carp and catfish breeding and culture (Ed. Langer, R. K., Kiran Dube and Reddy, A. K.), Central Institute of fisheries Education, Mumbai, pp 12-29.

Elias, Dadebo. 2000. Reproductive biology and feeding habits of the catfish Clarias gariepinus (Burchell) (Pisces: Clariidae) in Lake Awassa, Ethiopia. *Ethiopian Journal of Science.* 23(2): 213-246.

Hossler, F. E., Ruby, J. R. and McIlwain, T. D. 1979. The gill of the mullet, Mugil cephalus. I. Surface ultrastructure. *J. Exp. Zool.* 208, 379-398.

Johal, M. S. 1981. Food and feeding habits of some fishes of Punjab. *Vest. Cs. Spolac. Zool*; 45: 87-93.

Kant, Sashi and Shama Vohra. 1993. Role of algae as primary producer in fish production in Kashmir lakes. In: Advances in Limnology (Edited by H. R. Singh). Narendra Publishing House, Delhi, pp 79-86.

Lakra, W. S. and Singh, A. K. 2007. Exotic fish introduction in Indian waters-past experience and lesson for the future. *Fishing Chimes.* 27(1): 30-34.

Munshi, J. S. D., Ojha, J. Ghosh, T. K. and Roy P. K. and Mishra A. K. 1984. Scanning electron microscopic observations on the structures of gill rakers of some freshwater teleostean fishes. *Proc. Indian Natn. Sci. Acad. B.* 50(6), 549-554.

Ojha, J. Mishra A. K. and Munshi, J. S. D. 1987. Interspecific variations in the surface ultrastructure of the gills of freshwater mullets. *Japan J. ichthyol.* 33 (4), 388-393.

Piska, Ravi Shankar. 1999. Fisheries and Aquaculture, Lahari Publications, Hyderabad, pp. 452.

Reddy, P. B. S., 1980. Food and feeding habits of _Channa punctata_ (Bloch) from Guntur. _Indian J. Fish_; 27 (1&2): 123-129.

Routray P; P. Kumaraiah and N. M. Chakraborty. 2001. Some aspects of fishery biology and conservation of a peninsular carp, _Puntius pulchellus (Day)_. _Fishing Chimes_. 21(5): 53-55.

Saikia, S. K. and Das, D. N. 2008. Feeding ecology of common carp (_Cyprinus carpio L.)_ in a rice-fish culture system of the Apatani Plateau (Arunachal Pradesh, India). _Aquat. Ecol._ 43(2): 559-568.

Sakhare, V. B. 2010. Food and feeding habits of common carp, _Cyprinus carpio_ (Linn.). _Fishing Chimes_. 30 (1): 180-182.

Selvaraj, C., V. K. Murugesan and V. K. Unnithan. 1997. Ecology-based fisheries management in Aliyar reservoir. Central Inland Fisheries Research Institute, Barrackpore (West Bengal). Special publication No. 72.

Siva Reddy, Y., 1985. The biology of a few common fishes from Hussainsagar Lake, Hyderabad, India-Ph. D. Dissertation, Andhra University, Waltair.

Sugunan, V. V. 1995. Reservoir Fisheries of India. FAO Fisheries Technical Paper No. 345. Daya Publishing House, Delhi.

Sundararaj, V. and Srikrishnadhas, B. 2000. Cultivable aquatic organisms, Narendra Publishing House, Delhi, pp. 165.

Qayyum, A. and Qasim, S. Z., 1964. Studies on the biology of some freshwater fishes. Part 1. _Ophiocephalus punctatus_ Bloch. _J. Bombay Nat. Hist. Soc._ 61(1): 74-98.

Chapter 14
Management of Fish Farms

The fish farming cycle consists of production of fish seed by controlled breeding (hypophysation technique), operation of hatcheries and thereafter the rearing of spawn to fry, fingerling size and ultimately raising of the table size fish or brood fish. In this cycle of fish farming, the most important stage is the raising of seed. The tender baby fish (about three day old) is independent enough to lead its own life, but needs a conducive and congenial environment with plenty of natural food for its growth and survival. Thus, the processes of rearing of the spawn, measuring 5-7 mm in size, to 20-25 mm size (fry) in the small water bodies is known as nursery pond management.

Nursery Pond Management

The guiding principle of nursery management is to provide the young spawn sufficient nutrition, risk free living and a proper environment. Nursery ponds provide suitable ecological conditions and protection from destructive and predatory animals, and enriched nutrition, thereby ensuring high rate of survival and growth. The management of nursery pond includes the clearing of algal blooms and weeds, eradication of unwanted animals by poisons, liming, fertilizing, releasing spawn and their nursing, supplying adequate artificial feeds and harvesting fry for transference.

Pond Preparation

A nursery pond can be either earthen or cemented. Any small seasonal pond can be used as a nursery pond. The ponds are rectangular in shape, with an area of 200-400 m² each. The physical texture of the pond soil should be loamy or clay-loam so that water retentivity is good. The pond bottom should be free from excessive decomposing matter. Good productive soils have soil pH: 6.5-7.5, available Nitrogen: 30-50 mg/100g, available Phosphrous: 6-16.1 mg/100g and organic carbon: 1-2 per cent. The water should have turbidity below 20 ppm, pH 7-8.2; total alkalinity 75-150ppm and dissolved oxygen 4-10ppm.

Water should be available for at least three months to maintain a steady water level between 0.75 to 1.0m. Therefore, a good water source having ample supply of water, especially in late summer, for raising spawn of early breeding, is essential.

The nursery pond management operations start right from the months of February and March. Perennial ponds harbour predators and minnows; it is, therefore, always better to dewater such ponds and expose them for drying. Drying of such water bodies in the summer helps mineralization and removal of excess of reducing gases. In case of excess mud, the top layer above 10-15 cm, should be removed and can be used for strengthening the bundh which may be used for growing vegetables. Moreover, during the rains its manorial value becomes available to the pond and adds to its productivity. Minor repairs pertaining to the inlet-outlet are to be attended to during desilting time.

Eradication of Aquatic Weeds and Predators

Vide details given in Chapters 15 and 16.

Water Supply

Water from any source can be used for refilling nursery ponds after passing it through a fine sieve to get rid of predators and insects in any stages. Generally, nursery ponds are watered 10-15 days before stocking, initially to 60 cm, and subsequently raising it up to 100-120 cm.

Liming

Lime @ 300 kg/ha is applied after two weeks of eradication of predators and weed fishes. The advantages of liming in pond are numerous. In general, liming enhances pond productivity and improves its sanitation. It is both prophylactic and therapeutic. Specific advantages of lime are listed below:

1. Kills pond bacteria, fish parasites and their intermediate life –history stages; hence, especially efficacious in a pond where there has been outbreak of infectious disease.

2. Build up alkaline reserve and effectively stops fluctuations of pH by its buffering action.

3. Renders acidic waters usable for aquaculture by raising their pH to alkaline levels.

4. Neutralizes iron compounds which are undesirable to pond biota including fish.

5. Improves pond soil quality by promoting mineralization.

6. Precipitates excess of dissolved organic matter and thus reduces chances of oxygen depletion.

7. Acts as a general pond disinfectant for maintenance of pond hygiene.

Manuring of Pond

Manuring is done with the objective of encouraging of plankton, particularly animalcules which form the natural food of the spawn. Nurseries are thus manured with only cow dung (organic manure) @ 10,000 kg/ha about 15 days before the anticipated date of stocking by broadcasting all over the pond. In ponds where mahua oil cake is used as piscicide earlier, the dose of cow dung nay be reduced to half.

Stocking

If piscicide is earlier used for preparation of the pond, it is essential to ascertain its complete detoxification before stocking. This may be conveniently done under field conditions by fixing a 'hapa' in the pond and releasing some spawn in it. Comfortable behaviour of spawn for about 24 hours confirms complete detoxification. Nursery ponds are stocked with about 3-4 days old spawn usually in the morning hours after about 5 weeks of piscicide application, if done. The moderate rate of stocking may be 25-30 lakh/ha.

Supplementary Feeding

Even after manuring, it is difficult to maintain the desired level of natural food for the growing fry in the pond. Hence it is essential to provide supplementary feed from the outside. A mixture of finely powdered groundnut/mustard oil cake and rice bran/polish in equal proportion by weight is supplied to the fry. Other feed items that can alternatively be used are mixture of powdered aquatic insects, prawn and cow peas or fish meal and groundnut oil cake etc. It is recommended that cobalt chloride or manganese sulphate @ 0.01 mg/day/spawn may be added to the feed. Addition of yeast increases survival of fry. Feed may be broadcast all over the pond once daily in morning hours commencing from the day of stocking. Feeding may be stopped a day earlier to harvesting. The generally recommended feeding schedule is as below:

Period	Rate of Feeding per Day	Approximate Quantity per One Lake of Spawn/Fry
1st to 5th day after stocking	4 times the initial total weight of spawn stocked	0.56 kg
6th to 12th day after stocking	8 times the initial total weight of spawn stocked	1.12 kg
13th day	No feeding	-
14th day	Harvesting	-

Harvesting of Fry

On 14th day of stocking fry grows to about 25-30 mm size. They are harvested with the fine meshed (1.5 mm) drag net in the cool morning hours avoiding the cloudy days.

During a normal breeding season lasting about 3 months, 3-4 crops can be raised from the same nursery pond. The fry are then raised to fingerlings in rearing ponds.

Rearing Pond Management

The rearing pond must be located near the nursery and stocking ponds, so that the transportation of fry would be easy. The fry are raised to fingerlings in the rearing ponds. It require a period of about 3 months. The package developed by CIFRI for healthy growth and better survival of fingerlings is described below:

Pond Selection

Rearing ponds may be 0.05 ha to 0.1 ha in area. They may be rectangular in shape with water depth ranging from 1.5 to 2.0 meter. Seasonal ponds are often preferable to perennial ones. Care should be taken to prevent entry of ducks inside the rearing ponds.

Weed Eradication

Vide details given in Chapter 15.

Liming

Liming the pond @ 250-300 kg/ha in 3 equal monthly installments is recommended. The first dose need be applied one week earlier to stocking.

Pond Fertilization

Provision of natural food can be ensured to some extent in the pond through manuring and fertilization at fortnightly intervals. Organic manure (cow-dung) @ 2500 kg/ha is generally applied in 4 equal installments. While the initial dose is applied about a fortnight before the stocking, the subsequent ones are used on monthly intervals. However, when the ponds are earlier treated with mahua oil cake the first installment of cow dung can be dispensed.

Inorganic fertilizers like urea@ 100 kg/ha or ammonium sulphate @ 200 kg/ha and single super phosphate @ 100 kg/ha or triple superphosphate @ 35 kg/ha may be applied in 3 equal installments during the rearing period. The first installment of inorganic fertilizers is given on second day of stocking and thereafter at monthly intervals, alternating with organic manures. Manuring and fertilization need be suspended if algal blooms or any other adverse conditions appear in the pond.

Fry Stocking

After testing for complete detoxification, fry of 25-30 mm size are stocked in various combinations at densities ranging from 2 to 3 lskh/ha in any of the following ratios:

Species	Ratio
Catla + Rohu + Mrigal	3:5:2
Silver carp + Grass carp	1:1
Catla + Rohu + Grass carp + Mrigal	4:3:1.5:1.5
Silver carp + Grass carp + Common carp +Rohu	3:1.5:2.5:3

Supplementary Feeding

Supplementary feeding consists of a mixture of groundnut/ mustard oil cake and rice bran at 1:1 ratio by weight in powder form, and it is given every day in the pond during the morning hours. The feeding schedule, shown below, may be followed for 3 months of the rearing period.

Period	Quantity of Feed/Day/Lakh of Fry
First month	6 kg
Second month	10 kg
Third month	15 kg

The artificial feeds may be stopped when the pond water turns thick green or bloom develops. Also floating aquatic vegetation such as Azolla, Lemna and Spirodella should be introduced as it cuts off light penetration and controls algal bloom, and also serves as good feed for grass carp seed.

Fingerling Stocking

The growth and well being of growing fry need to be checked by sample netting, at least at monthly intervals. Healthy fingerlings of 100-150 mm size are obtained in three months rearing period. Supplementary feeding is stopped a day before the date of catching. Harvesting is done during cool morning hours by repeated drag netting. Survival of about 70-90 per cent (average 80 per cent) is generally obtained in about 3 months rearing.

Production/Stocking Pond Management

Pond Selection

The ideal size of the production pond for profitable fish culture is 0.1 to 2.0 ha. In this pond fingerlings are stocked and grown for about one year or until they attain a marketable size of about 1 kg each.

Liming and Fertilization

If the soil of the stocking pond is found to be acidic, it has to be treated with quicklime to bring the soil pH to alkaline condition, which is ideal for fish production. Lime is applied when all the water is drained out and the bottom is well dried, exposed to the hot sun for about 15 days. Lime may kills parasites of fish at bottom. The dosage of quicklime depends on the pH of the pond soil as given below:

pH	Nature of Soil	Quantity of Quick Time to be Applied (kg/ha)
4.5-5.0	Strong acidity	2000
5.0-6.5	Medium acidity	1000
6.5-7.5	Neutral	500
7.5-8.5	Medium alkalinity	200
8.5-9.5	Strong alkalinity	No application

The quantity may however be increased by 50 per cent, if the bottom is clayed and reduced by 50 per cent for a sandy bottom. The quicklime is either sprinkled or spread on the pond water in the form of a paste to act as an antiseptic and also to neutralize the toxic effect of old organic deposits at the bottom. It also stabilizes the pH of water at a slightly alkaline level, which enhances the growth of phytoplankton and fish apart from increasing the calcium content of water. Further, it increases not only the bicarbonate content of the pond but also counteracts poisonous effects of ions like magnesium and sodium.

Every pond soil is likely to have some amount of stored nutrients, which are released in the water slowly for its productivity. Compared to this stage of nutrients, the quantity of water borne nutrient is insignificant. Though the pond soil is capable of releasing nutrients periodically, at times, it may stop releasing nutrients. Under these conditions, fertilizing the pond is inevitable. Application of fertilizer is basically responsible for the development of primary producers, *viz.* phytoplankton on which zooplankton feed. The culturable fish in turn depend either on the latter or on the former directly, based on the type of their feeding habits.

After 15 days of liming, the fertilization is to be done in order to develop the fish food organisms. Any production pond for fish culture is to be fertilized by both organic and chemical fertilizers.

Pond Fertilization

Depending on the quality of the soil characteristics such as pH, organic carbon content, the quantity of organic manure should be determined. If there is less organic carbon manure, cow dung for the stocking pond is applied at the rate of 20-30 t/ha and if there is more organic carbon reserve in the soil, the manure may be reduced by 50 per cent. However, poultry- manure at the rate of 5000 kg/ha is known to enhance zooplankton production. The first installment should be one-sixth of the total quantity a fortnight, before stocking with fingerlings and the rest in 10 equal monthly installments. If mahua oil cake has already been used, the first installment of organic manure may be reduced by 25 per cent.

The use of chemical fertilizers will vary according to the concentration of phosphorus and nitrogen in the soil. It is reported that hard water need more nitrates and soft water more phosphates. The standard combination of NPK as 18:10:4 is generally recommended. For a production pond of medium fertile soil, urea at the rate of 200-kg/ha/yr or ammonium sulphate at the rate of 450 kg/ha/yr, superphosphate at 250 kg/ha/yr and muriate of potassium chloride at 40/kg/ha/yr should be applied in equal installments, alternating with organic fertilization. However, application of fertilizers should be suspended, if and when, algal blooms are noticed. Every care should be taken while using ammonium fertilizers.

Stocking

After 15 days of initial fertilization fingerlings are stocked in production ponds. Prior to stocking, the fingerlings if brought from outside, are first quarantined and acclimated with the water of the production ponds. The fingerlings are then given a dip treatment with 2 per cent potassium permagnate solution in order to safeguard them from parasitic infections. Though the stocking density of fish fingerlings varies in different culture practices, it is generally based on the principle that an area of $1m^2$ is needed for a fish to grow and attain an average weight of 1 kg/yr. Considering this, the stocking density of fingerlings in a production pond may be about 10,000/ha. However, depending on the fertility of water, species and size of fish cultured, culture period and type of culture practice, the stocking density may vary from 2000 to 10,000/ha.

Feeding

The method of feeding in a production pond is similar to that for rearing pond described earlier. However, depending on the nature of fish grown and their feeding habits, the materials for feeding and methods of feeding will differ to some extent.

Harvesting of Fish

At the time of harvesting, the water is gradually drained and the fish are harvested. Based on the need and marketability, partial or complete harvesting may be done. It is advisable to harvest the fish in the early morning or in the evening to keep the harvested fish in healthy condition for marketing.

Chapter 15
Aquatic Weeds and their Control

Jethrotull (1731) used weed first time in his famous book 'Horse Hoeing Husbandry'. Aquatic weeds are those plants, which start in water and grow at least part of their life cycle in water. One of the major problems faced by fish farmers is to control unwanted and undesirable vegetation since its profusion these appropriates disproportionately in large quantities of soluble nutrients drastically curtailing their productive role to economy. They hamper the growth of fish, plankton, upsets the equilibrium of physico-chemical qualities, give shelter to undesirable fauna, insects and molluscs and fishes and obstruct netting etc. In a balanced ecosystem the water with a variety of flora and fauna dissolved and submerged plants and gases mainly O_2 remain under check through the interaction of various physical, chemical and biological factors operating both within and outside the water body. When this balance is disturbed rapid changes occur and growth of aquatic flora takes place.

Disadvantages of Weeds

1. Weeds check free movement of fish
2. They cause oxygen depletion and accumulation of carbon dioxide

3. Aquatic weeds are harmful as they consume nutrients of the pond

4. They obstruct the netting operations

5. Gases like hydrogen sulphide and methane are formed which are harmful to the fish.

6. Weeds increase the cost of labour and equipment.

7. Algal blooms choke the gills and spoil the water on rotting, and also silt the ponds.

Advantages of Aquatic Weeds

1. When present in limited quantity are useful and necessary for the ecology of the pond.

2. They form natural food of many fishes, and fertilize the pond when decayed.

3. They provide shade and shelter to many fish and oxygenate the water.

4. They reduce turbidity and provide spawning beds for fishes.

Types of Weeds

On the basis of habit and habitat, Lawrence (1955) has classified weeds as:

Floating Weeds

These weeds have their leaves freely floating on the surface of water, and roots hanging underneath. They are profusely found in wind-protected ponds. They are harmful as they shade the ponds. *e.g. Pistia, Eichornia, Lemna, Spirodela, Azolla,* and *Wolffia.*

Emergent Weeds

These are rotted in the bottom of the pond, but their leaves float on the water surface or rise above the water level. They prefer shallow parts and shores of the pond. *e.g. Nymphaea, Nelumbo,* and *Nymphoides.*

Submerged Weeds

These plants grow under the water surface and may or may not be rooted. Examples of submerged weeds are *Hydrilla, Najas,*

Vallisneria, and *Potamogeton* etc. Rootless plants are *Utricularia*, *Ceratophyllum* etc.

Marginal Weeds

These weeds are mostly rooted and infested in shallow foreshore areas of a water body. *e.g.* Typha, *Cyperus, Panicum, Phragmetes, Panicum, Colocasia, Thaimoea* and *Eclicharis*.

Mats and Scums

These weeds are floating on the surface. *e.g. Spirogyra, Pithophora* etc.

Algae

Algae that are dispersed through the waterbody. *e.g. Microcystis, Oscillatoria, Anabaena* etc.

On the basis of contact with soil, water and air, Mirashi (1957) classified the weeds into different six types *i.e.*,

Floating Hydrophytes

Contact with water and air only. *e.g. Ipomoea, Eichornia, Trapha, Hygorhiza* etc.

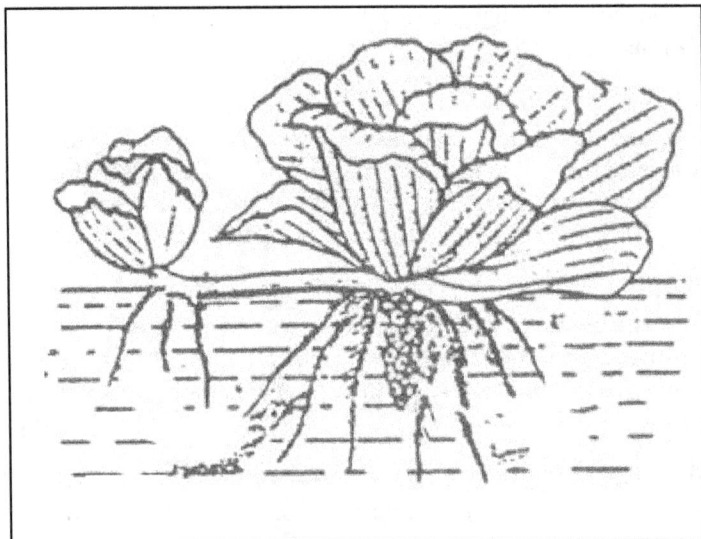

Figure 15.1: Pistia

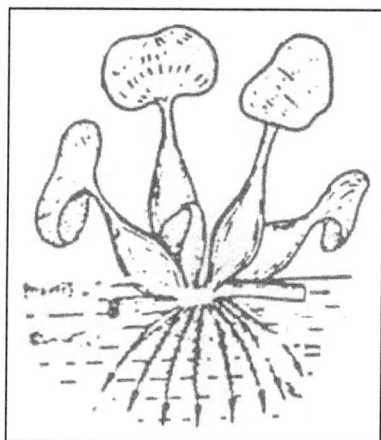

Figure 15.2: Eichhornia

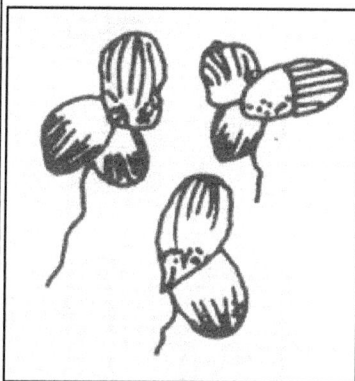

Figure 15.3: Lemna

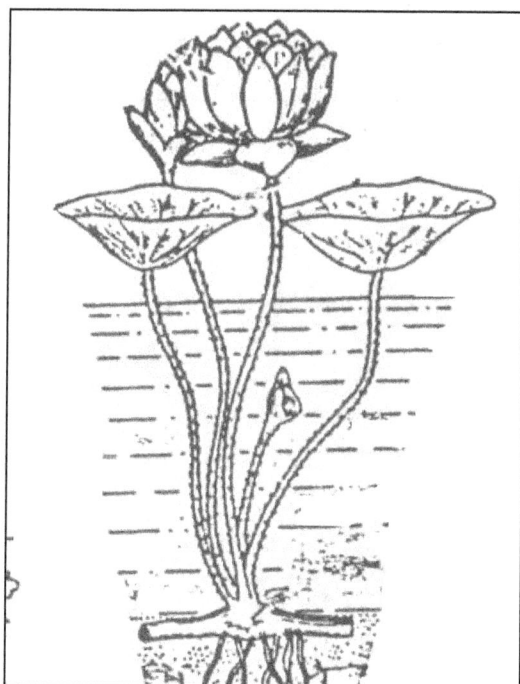

Figure 15.4: Nelumbo

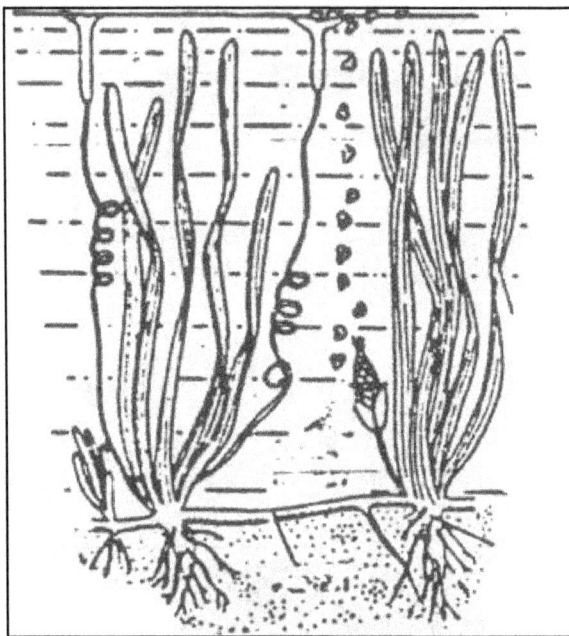

Figure 15.5: Vallisneria

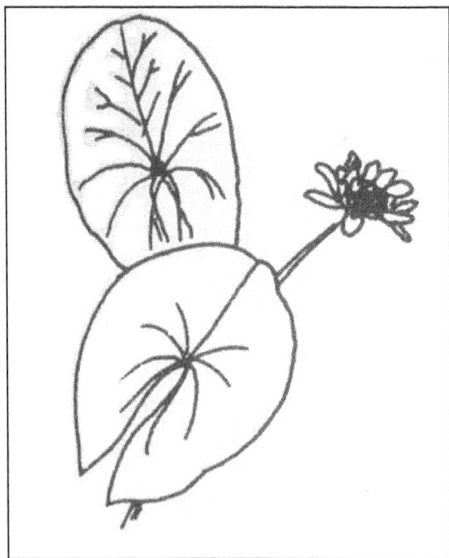

Figure 15.6: Nymphaea

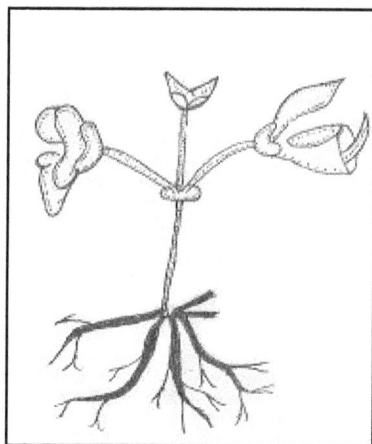

Figure 15.7: Utricularia

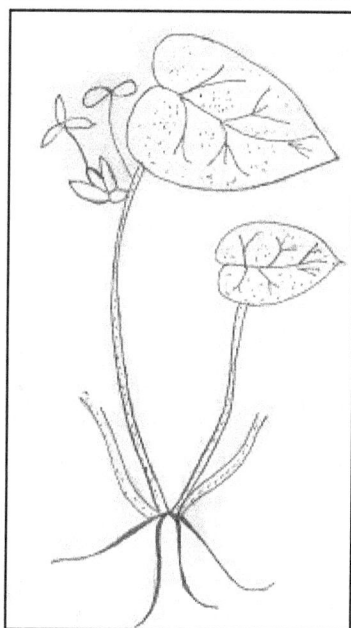

Figure 15.8: Nymphoides

Figure 15.9: Cyperus

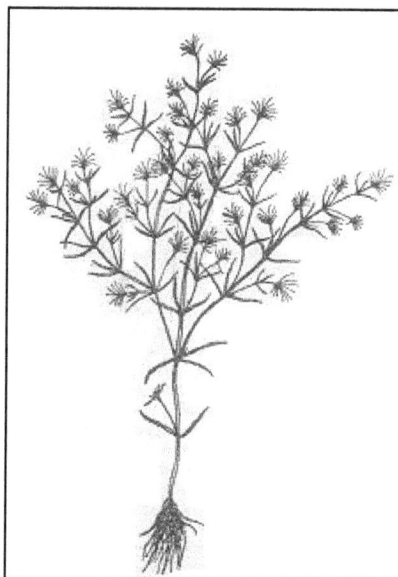

Figure 15.10: Najas minor

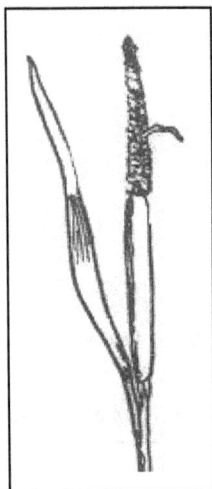

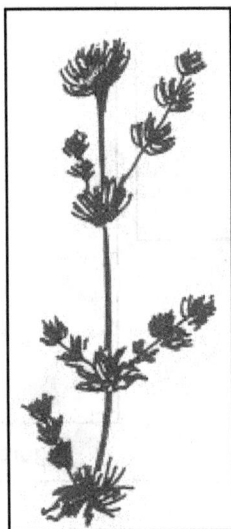

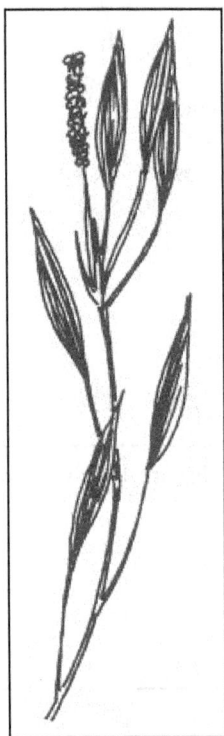

Figure 15.11:
Typha

Figure 15.12:
Ceratophyllum

Figure 15.13:
Potamogeton

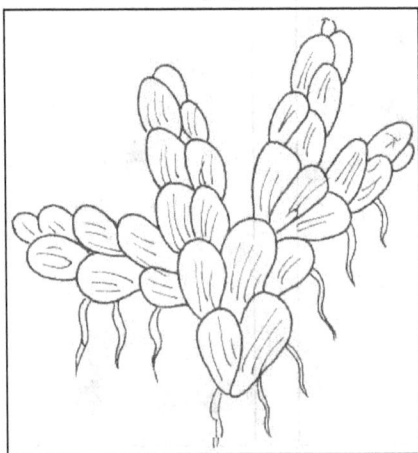

Figure 15.4: Azolla

Suspended Hydrophytes

They are rootless submerged hydrophytes. In contact with water only. *e.g. Ceratophyllum, Utricularia* etc.

Anchored Submerged Hydrophytes

Most part is in contact with soil and water only. *e.g. Najas, Hydrilla, Vallisneria, Blyxa* etc.

Anchored Hydrophytes with Floating

They contact with soil, water and air. *e.g. Saggitaria, Ottelia, Nymphaea, Aponogeton* etc.

Emergent Amphibious Hydrophytes

These weeds are able to live in land and in water. Their roots, lower portion of stem, in some cases even lower leaves usually submerged in water. *e.g. Sebinia, Limnophyton, Hygrephya* etc.

Wetland Hydrophytes

They are rooted in soil, usually saturated with water at least in the early part of their life. *e.g. Polygonum, Commelina, Ammania, Malachra* etc.

According to Hickling (1962), aquatic weeds are of two types *i.e.*, hard plants and soft plants.

Weed Control

Manual Methods

It is a very old practice for control of weeds. Human energy is used with appropriate tools *i.e.*, hand cutting of hand pulling of aquatic weeds.

Mechanical Methods

These involve physical forces to remove the aquatic weeds or alter the aquatic environment. So that the plant cannot become established or survive, if already present in the waterbody. When a machine is used for physical control of weeds, method is known as mechanical methods. There are number of specialized equipments for crushing, dredging, mowing, harvesting and lifting of aquatic weeds. This method provides efficient weed control.

Advantages of Mechanical Control of Weeds

1. In developing countries like India, the herbicide availability is limited. Therefore, mechanical control is feasible.

2. Manpower is cheaper in developing countries.

3. Mechanical method never causes environmental pollution.

4. There is no risk of non-target organic and non-functioning of ecosystem.

5. This method is non-selected and there is less chances of development of new weed.

6. Periodical removal of aquatic weeds from water bodies remove huge quantities of excess nutrients from waterbody, which slows down the growth of other weed spp. while in case of chemical control the weeds decompose *in situ*.

7. It gives immediate result and there is no time lag.

8. Man may use the harvested weeds in many ways.

Limitations of Mechanical Controls

1. Moving the lawn weeds regrow fast from propogates. Even digging of canal bottom is not a success because weeds re-grow and come to the level before, when there is water filled again. Therefore, it should be well-timed and continuous process and supplemented with biological and chemical methods.

2. It allows spread of weeds in new water bodies by the way of plant fragments and reproduce vegetatively.

3. By the repeated removal of weeds from a water body, the plant nutrients deplete which are required for the growth of phytoplankton, which is a primary food of fish. But this is not a serious problem due to presence of periodic source of farm or city water and fertilizers may be applied for the growth of phytoplankton.

The various mechanical methods of weed control include chaining, dreding, use of weed witch etc.

Chemical Control

Chemical control technology is highly effective but requires proper and careful application to avoid water contamination, which results in toxicity to non target plants and aquatic fauna. Several chemical weedicides are now available for the control of aquatic weeds, but they have to be used carefully to prevent adverse effect on the fishes in water. The chemical is to be selected and used in such a way that:

☆ It should be cheap and easily available.

☆ Non-toxic to fish and man.

☆ Should not pollute the water, and

☆ Should not involve the use of special and costly equipment.

Of all the floating weeds, the water hyacinth, *Eichhornia crassipes*, is the most common and of greatest nuisance value. It can be successfully controlled by the chemical 2,4-dichlorophenoxy acetic acid (2,4-D). This chemical is applied at the rate of 4.5 to 6.5 kg/ha, and has harmful effects on fish. It kills the plants by deadening the leaves and may also be absorbed by the roots destroying them. Pistia and Lemna can also be treated with 2,4-D. Another weedicids is Taficide-80 (2,4-D sodium salt 80 per cent) and when used at the rate of 4-6 kg/ha in an aqueous solution of 1-1.5 concentration in combination with 0.25 per cent solution of surf is reported to kill the water hyacinth completely. Simazine cethyl-amino-s- triazine is also an effective weedicide.7.5-10lbs of chloroxone in 100-200 gallons of water can be sprayed on an acre of water. This chemical kills the plants in 2-3 weeks but is harmless to fish and the operator. It has to be used on clear sunny days, as rain washed off the chemical from the leaves. *Pistia* and *Lemna* can also be killed in 10 days by using 1 per cent solution of sodium arsenite (35lbs/acre), but this is poisonous and precautionary measures must be taken. *Pistia* is also reported to be killed by using ammonia at a concentration of 40 ppm N.

Marginal weeds like *Typha, Colocasia, Sagittaria* and grasses grow along the pond margin and harbour predatory insects. Although the most effective method is to cut them by manual labour, they can be controlled by chemical weedicides. Young *Cyperus* and *Colocasia* are totally killed with 1-1.5 per cent aqueous solution of Tacifide-80 along with detergent surf.

Emergent weeds such as water lilies and lotous are the common rooted plants with leaves emerging on the surface. Pulling them out with manual labour or surface treatment with chemicals is of little use in eradicating them. Use of Taficide-80 is reported to be effective in controlling Nelumbo. Lily plants get uprooted by treatment with 2,4-D sodium salt at 1.5 per cent concentration, along with 0.25-1 per cent detergent surf.

Submerged weeds like *Vallisneria, Potamageton, Najas, Ceratophyllum,* and *Hydrilla* are most trouble-some, and their removal by manual labour is tedious and costly. The submerged weeds are killed if the water is made turbid for a longtime. Copper sulphate in combination with ammonium is also found to be effective. Sodium aresenite at 5-6ppm is also found to be effective in killing submerged weeds. The inorganic fertilizers like superphosphates and urea have a toxic effect on these weeds. Experiments conducted at Central Inland Fisheries Research Institute (CIFRI) have shown that superphosphates at 500 ppm completely killed submerged weeds. *Hydrilla* was completely killed with urea at 250-300 ppm. But urea is harmful to the fish, as ammonia is released. Anhydrous ammonia is also effective in killing the sub-merged weeds, but it is harmful to the fish.

The filamentous algae rarely become a nuisance, but sometimes spirogyra and phytophora form extensive mats. Sodium arsenite is effective in killing them especially in warm weather. Algal blooms consisting of *Microcystis, Anabaena* and *Oscillatoria* are sometimes harmful by choking the gills of fish, by causing oxygen depletion or by rotting. They can be eradicated by treatment with copper sulphate. Simazine at a small dose of 0.5-1.0 ppm is also effective in controlling microcystis.

Biological Control

Biological control means employing natural enemies on target weed species so that the bioagents attack the weed plants and kill them. It may be accomplished by using herbivorous animals, predators, pathogens, and parasites.

Several selected species of fish can control certain hydrophytic weeds and established their usefulness for cleaning waterbodies. Chinese Grass carp (*Ctenopharyngodon idella*) is a purely herbivorous. It is adapted to a wide variety of situations and makes no demand on water quality. It thrives even at low oxygen (10,000 ppm) levels. Though the carp can devour large masses of hydrophytic weeds but the quantity of consumption depends upon water temperature. It eats anything at temperature lower than 4°C but in between 14°C and 18°C, it likes soft tissues of tender shoots. When the temperatures raised above 18°C, the carp swallows everything it mets such as submersed, emersed, and floating herbs. In China, some studies indicated that the fish consume 50-70 per cent of its body weight of water grasses everyday and 100 fish can eliminate the aquatic weeds in a few days, in a hectare of pond. It consumes submersed weeds such as hydrilla, chara, potamogeton, and ceratophylum. Less than 1.5 kg fish eat several times more than its body weight and gain weight @ 6 g/day. Being a quick consumer, every year it gains 2-3 kg body weight.

Common carp is an omnivore and eat submersed vascular weeds and filamentous algae but also feed on insect larvae and small aquatic fauna. Common carp is primarily a scavenger and scavenge around bottom muds of ponds. It uproot submersed weeds in search of rhizomes and roots of water plants. Ponds should be stocked with 100 to138 carps/ha for effective weed control. Other fishes like Silver carp (*Hypophthalmichthys molitrix*), Silver dollar fish (*Methynnis roosevelti Eigen*), Tilapia (*Tilapia* spp.), Tawes (*Puntius javanicus*), and Gold fish (*Carassius auratus*), are also useful in biological control of aquatic weeds.

Chapter 16

Fish Predators and their Control

Nursery ponds should be free from harmful fauna such as snakes, frogs, predatory fishes and weeds fishes which do not only prey upon the hatchlings and fry but compete for food, space and oxygen. In view of the harm caused by these undesirable fishes, their complete eradication from the pond before stocking is of utmost importance in scientific nursery pond management. The common predatory fishes found in ponds are *Channa spp, Clarias batrachus, Heteropneustes fossilis, Glossogobius giuris, Mystus* spp, *Ompak* spp, *Wallago attu, Pangasius pangasius* etc. The common weed fishes are *Puntius* spp., *Oxygaster* spp., *Ambassis* spp., *Aplocheilus* spp, *Esomus danricus, Amblypharyngodon mola, Colisa* spp. etc.

Control of Predatory and Unwanted Fishes

The control of predatory and unwanted fishes is done by draining the pond water or by application of fish toxicants. The fish toxicants are of plant origin, or chemicals, including pesticides. The toxicants are selected on the basis of their properties like effective minimum dose, least advance effect on pond biota, short duration of toxicity, and consumability of poisoned fish. Some of the common fish toxicants applied in nursery ponds are given below:

Mahua Oil Cake

The plant *Bassia latifolia* is commonly known as mahua in India. The oil cake of this plant is extensively used as fish toxicant. It contains about 4-6 per cent saponin. All unwanted fishes are killed at a dose of 200-250 mg/lit within 6-10 hours. The toxic effect in water lasts up to 15 days. After its initial toxicity is used up, to kill the fish, it acts as a fertilizer and greatly increases the phytoplankton standing crop of the ponds. The killed fishes are fit for human consumption. The quantity required is calculated on the basis of the volume of the water of the pond, 2000-2500 kg/ha oil cake is normally used. The required quantity of cake is powdered, soaked in water and broadcasted uniformally over water surface, repeated stirring of water, to ensure proper mixing of oil cake. This could be done with the help of drag netting. Affected fishes are removed from the pond.

Tea Seed Cake

Seed cake of tea (*Camellia sinensis*) may prove to be a substitute for mahua oil cake wherever commercially available. It can also be used @ 525-675 kg/ha in pond of wild fish, tadpoles and insects. The toxicity lasts for 10-12 days. The killed fishes are fit for human consumption and the tea seed cake also acts as a fertilizer ultimately.

Other Plant Derivatives

Other fish toxicants of plant origin like root powder of *Belanitis roxburhii*, unripe fruit of *Randia dumetrum*, seed of *Croton tigilium*, root of *Milletia pachycarpa*, roots of *Derris latifolia*, bark powder of *Walsura piscidia*, bark of *Xanthoxylum alatum*, entire plants of *Eupatorium odoratum*, bark of *Anagallis arvensis*, Latex of *Euphorbia royleana*, and fruit pulp of *Sapindus mucorosa* have been found to be effective under laboratory conditions.

Ammonia

Anhydrous ammonia @ 20-25 ppm has been found as an effective fish toxicant. The killed fishes are safely consumable. The cost of ammonia as a fish toxicant is off set to some extent by its fertilizer value which has been estimated to be about 36 per cent of the cost of ammonia applied. Ammonia acts as herbicide as well. Toxicity of ammonia in water lasts for about 5-6 weeks.

Anhydrous ammonia from a cylinder is introduced into the water through a hose and a 1.2m long G. I. pipe applicator with delivery holes. The applicator is suspended from a boat or held in position by long ropes by two persons standing on opposite bank. The cylinder is partly immersed in water near the shore to prevent excessive cooling in condensation of ice. Ammonia being lighter than water, the applicator is kept as far below the water level as necessary to effect the bottom dwellers. However, the applicator should be kept well above the bottom soil to prevent loss of gas from absorption by the soil. Its efficacy also depends on the pH of water, effect being quicker with increasing pH.

Application of Quicklime

It can be used in place of a toxicant to kill unwanted fishes, insects and tadpoles besides ameliorating the environment of the ponds. For this purpose a dose of 900-1050 kg/ha of quick lime is necessary if the level of water is low in the pond. If the pond is full, the dose of quick lime should be increased to 1575 to 2250 kg/ha. This is also a disinfectant.

Bleaching Powder

Bleaching powder (Calcium hypochlorite) as fish toxicant has been found to be effective in 3-4 hours at 25-30 ppm. Its toxicity lasts for about 7-8 days in the pond. It also possesses disinfecting effects besides oxidizing the decomposing matter on the pond bottom. In view of limited supply of mahua oil cake, bleaching powder is an effective substitute with easy availability and lower cost. The powder is dissolved in water and made into the form of a slurry. The solution is sprayed on the water surface immediately and the water stirred for thorough mixing. Distressed or killed fishes are removed by subsequent repeated netting and are fit for human consumption.

Control of Insects

Aquatic insects are less than 4 per cent of the total existing insect fauna of the world. These have high fecundity leading to build-up a large population in nursery ponds. Most of the aquatic insects, either in their larval and/or adult stages, not only prey directly upon spawn and fry of cultivable fishes but also compete with carp fry for food and space. So, it becomes very important for a fish farmer to eradicate them.

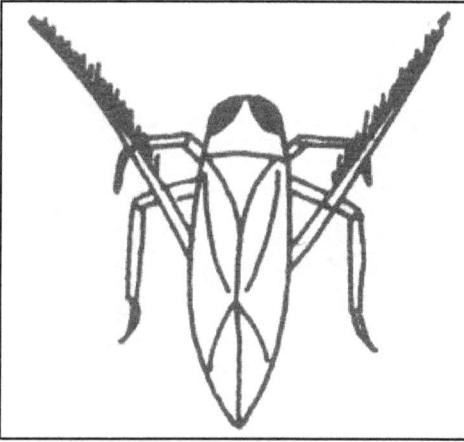

Figure 16.1: Backswimmer (Notonecta)

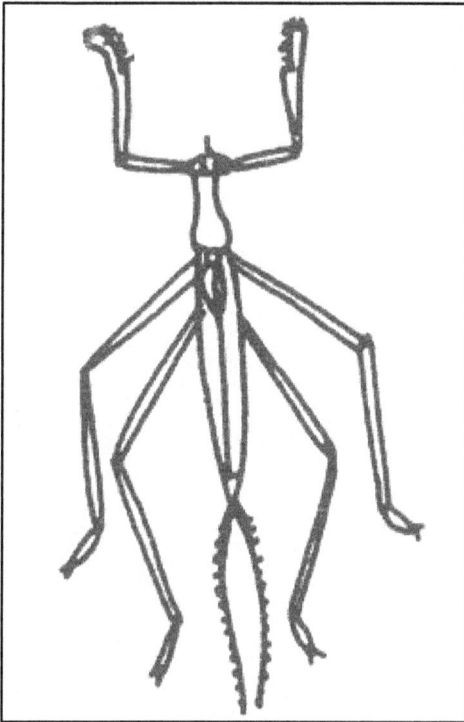

Figure 16.2: Water Stick Insect (Ranatra)

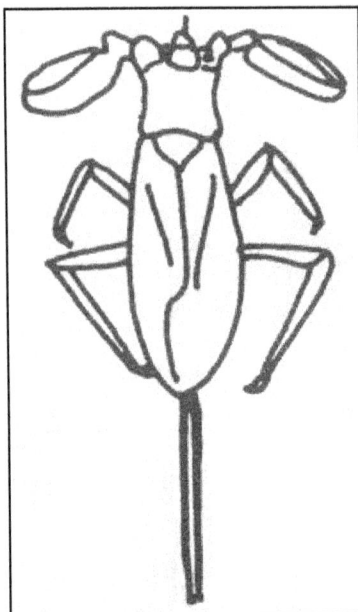

Figure 16.3: Water Scorpion

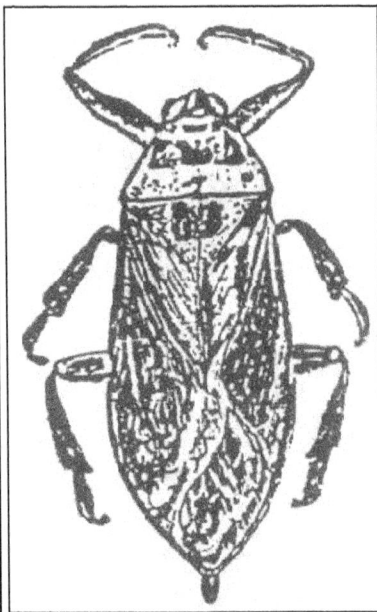

Figure 16.4: Belostoma

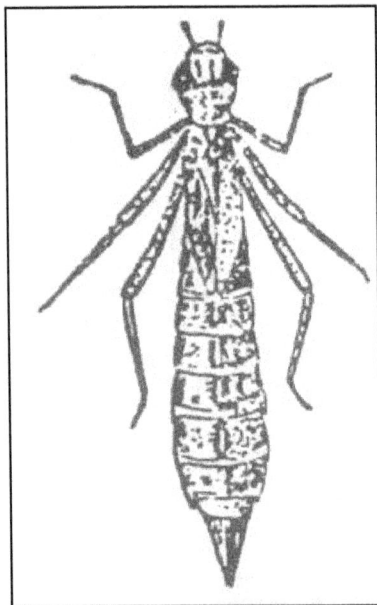

Figure 16.5: Drgonfly Nymph

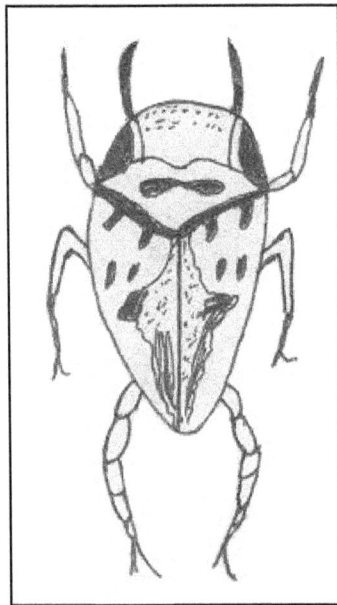

Figure 16.6: Laccophilus

Figure 16.7: Anisops　　**Figure 16.8: Water Beetle (Dytiscus)**

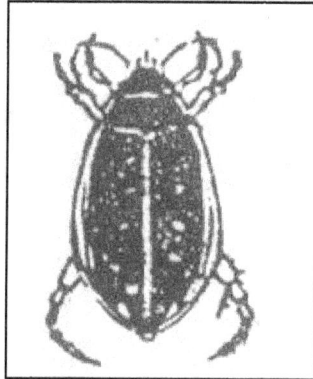

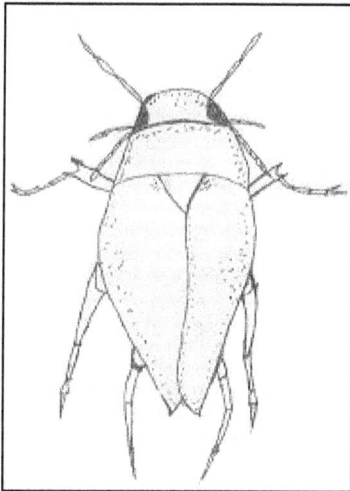

Figure 16.9:　　　　**Figure 16.10:**
Hydrophilus　　　　**Laccotrphes**

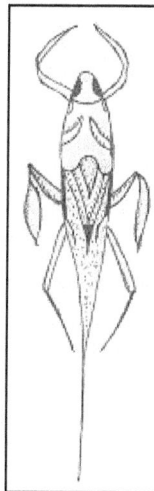

The most commonly occurring aquatic insect groups are:

Coleoptera (Beetles)

Predaceous diving beetle (*cybister*), water scavenger beetle (*Sternolophus*), and whirling beetle (*Gyrinus*) etc.

Hemiptera (Bugs)

Back swimmers (*Anisops*), Giant water bug (*Belestoma*), water scorpion (*Ranatra*), water stick insect (*Laccotrephes*)

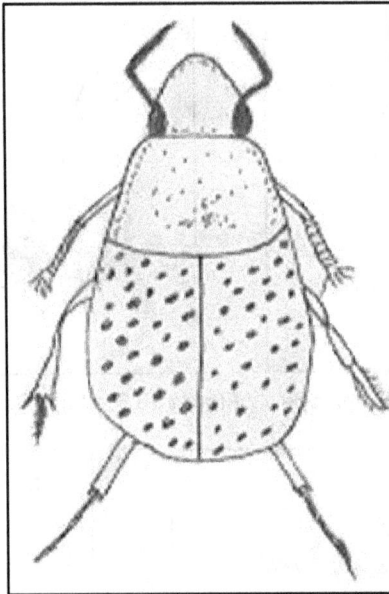

Figure 16.11: Sternolophus

Odonata

Dragonfly nymphs etc.

Since complete eradication of insects by means of seining is not possible, selective treatments are necessary. The commonly used methods to control insects in fish farms are as follows:

Use of Soap Oil Emulsion

CIFRI recommended to spray a soap oil emulsion on the surface of the water for killing insects which come up to the surface for respiration. An emulsin f any cheap washing soap and vegetable oil at the rate of 18 kg soap: 56 kg oil per hectare is prepared by heating the mixture for a short while. The emulsion is applied 12-24 hours before stocking by uniformly broadcasting over the water surface of the nursery pond. Care is taken to keep the uniform film undisturbed for a couple of hours. Hence, calm dry days are chosen for the application.

Soap can be substituted by Teepol B-300 (a synthetic detergent) in the above said emulsion. The recommended dose of Teepol is 560 ml to be emulsified with 56 kg of vegetable oil.

Use of Kerosene Oil

Spraying of kerosene oil over the surface of water @ 80-100 liters/ha has also been found to be useful by West Bengal State Fisheries Department in killing the aquatic insects in nursery ponds.

Use of Light Speed Diesel (LSD)

Another treatment recommended by Maharashtra State Fisheries Department is the spraying an emulsion prepared by mixing light speed diesel oil (1 liter), emulsifier Hyoxid 1011 (0.75 ml) and water (40 ml) at the rate of 1040.75 ml per 200 square meter of water surface.

Use of Turpentine Oil

Turpentine oil @ 75 liters/ha when sprayed on the water surface has been reported to kill the insects completely.

Use of Chemicals

The chemicals used for eradication of insects are Benzene hexachloride (0.1 ppm), Endrin (0.001 ppm), Dieldrin (0.5ppm), Gammaxene (0.5-1.0ppm) and quick lime (2-4 ppm).

Control of Other Predators

To control frogs and their tadpoles, snakes, birds etc. suitable measures have to be followed. By applying quicklime, the eggs of frogs can be killed. However, for avoiding the entry of adult frogs and snakes into fishponds, wire traps, nets or dragnets may be used. Shooting, scaring or netting invariably controls birds like herons and cranes.

Chapter 17

Cultivable Fishes

Indian Major Carps

Catla catla

This is one of the Indian major carps. The body is deep and moderately compressed. Head is large and very conspicuous. Eyes are large and situated in the anterior half of the head. Mouth is upturned and large. Lips are nonfringed. Barbels are absent. Dorsal fin is broad with 14-16 rays, which are branched. Body is greenish dorsally and silvery on sides and ventrally. Fins are dark and caudal fin is deeply forked. Scales are big and cycloid type. Lateral line is complete with 0-43 scales. Dorsal fin is inserted above tip of pectoral fin, and has 17-19 rays. It feeds on zooplankton of water surface using large gill rakers. It is a fast growing species among the Indian major carps. Catla grows to a length up to 45 cm, weighing more than a kilogram in one year and attains 2.2 kg and 6.5 kg weight at the end of second and third years respectively. The maximum size of catla has been recorded as 45 kg. It matures in the second year. It breeds naturally in rivers during the rainy season, though artificial propagation by hypophysation is possible. Its eggs are round, non-adhesive and nonfloating with diameter of 5.3 to 6.5 mm. The fecundity rate of catla varies from 80,000 to lakh eggs/kg of body weight. It is distributed throughout India, Pakistan, Nepal, Bangladesh and Thailand.

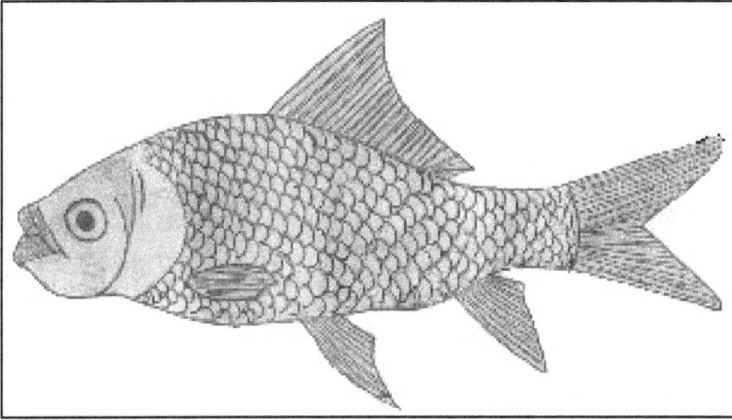

Figure 17.1: *Catla catla*

Labeo rohita

This is one of the Indian major carps, commonly known as rohu. Body is compressed and fusiform. Head is depressed and is produced into a short and blunt snout. It bears a subterminal fringe-lipped mouth bounded by fleshy upper and lower lips. It also contains paired nostrils and paired eyes. A pair of filamentous barbels arises from upper lip. Small tubercles cover the snout, which is oblong, depressed, swollen and projecting beyond the jaws. Lateral line is distinct. The colour of the fish is bluish –black along the back, reddish black along the sides and silvery in the abdominal area. Dorsal, anal, caudal and paired pectoral fin with soft fin rays. Caudal fin is forked.

It is a column feeder. It feeds on phytoplankton, plant debris or decaying debris of aquatic plants; however the young feed on zooplankton. Next to catla, it is known for its fast growth. It attains a length of 35-45 cm, with a weight of 0.7 kg to 1kg in the first year. At the end of the second and third years, it reaches around 2 kg and 2.5 respectively. It attains maturity by the end of second year. It is capable of breeding in ponds after inducement by pituitary gland extract. In natural conditions, it spawns once in a year. But by induced breeding, it can breed twice in a year. The fecundity rate of rohu varies from 2 to 5 lakh eggs/kg of body weight. Eggs are non-floating, on adhesive, and reddish in colour with diameter of 5 mm. It is distributed throughout India, Pakistan, Bangladesh, Nepal and Burma.

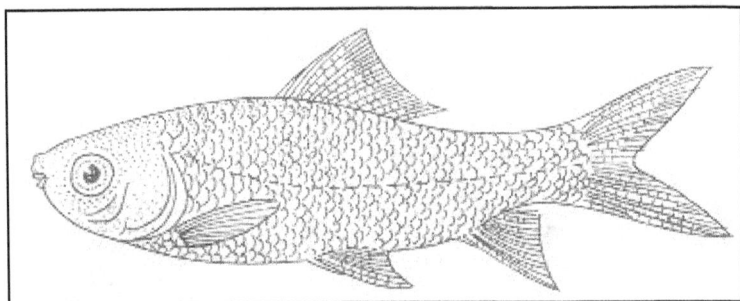

Figure 17.2: *Labeo rohita*

Cirrhinus mrigala

The body is narrow and linear. Head is small and snout blunt. The mouth is terminal. Lips are thin and nonfringed. Dorsal fin with 12-13 branched rays. Body is bright silvery in colour. The tip of the head is flattened and upper jaw is fringed. Lateral line scales are 40-45 in number. It feeds on plant and animal materials in the pond bottom. It consumes more quantities of decaying organic and vegetable matter. It grows to a length of 38.4 cm recording 1 kg in the first year. During the second and third years, its growth is around 1.5 kg and 3.5 kg respectively. It breeds in the natural waters by the end of the second year following maturity. It also responds to induced breeding technique in confined waters. It fecundity varies from 1,44,000 to 1,52,000 eggs/kg of body weight. It is distributed in India, Bangladesh, Nepal, Pakistan, Burma and East Indies.

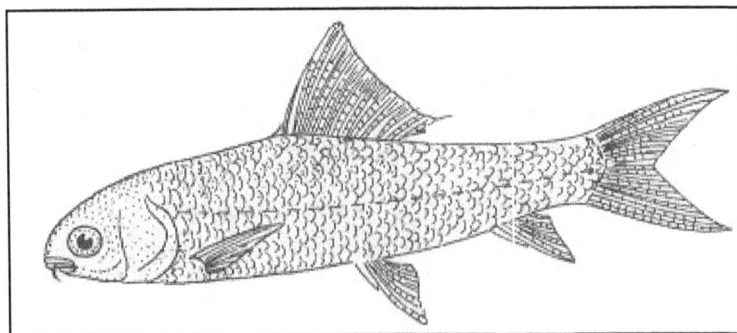

Figure 17.3: *Cirrhinus mrigala*

Exotic Carps

Ctenopharyngodon idella

It was introduced in India in 1959 by importing from Hongkong and Japan. It has been also imported and propagated in Vietnam, Thailand, Malaysia, Sri Lanka, Burma, Philippines, Singapore, Yugoslavia, Hungary etc. The body is cylindrical and elongate. Head is broad and snout is rounded. Upper jaw slightly larger than the lower jaw. Mouth is horny. Barbels are absent. Abdomen rounded. Dorsal fin small with less than 9 rays medially placed. There is a notch on the snout. Scales are cycloid and moderate in size. Dorsal surface greenish or dark grey in colour and silvery on the belly. Normally it feeds on aquatic weeds. It needs a minimum food of 25 per cent of its total body weight daily. Its maximum daily feeding ration has been as high as eight times of its total weight. Its habit of feeding aquatic weeds is beneficial in biologically controlling the aquatic weeds and it also serves as a 'living green manuring machine' besides its own growth. In the first year, it grows to 1.5 kg and in the second and third year it recorded maximum growth of 4 and 7 kg respectively. There are specific cases, where in some grass carps have grown to 3 kg in one year. The fecundity of fish is 1,34,000 eggs/kg of body weight.

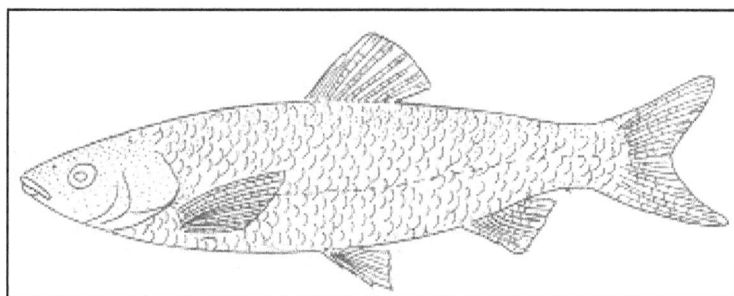

Figure 17.4: *Ctenopharyngodon idella*

Hypophthalmichthys molitrix

It is an exotic carp introduced in India during 1959 and suitable for culture in confined freshwaters along with Indian major carps. This has an oblong, slightly, laterally compressed body with a pointed head. Snout is bluntly rounded. Lower jaw slightly protruding, small eyes covered with adipose. Dorsal fin small with less than 9 rays and medially placed. Scales are very small about 110 to 124 in numbers on lateral line. It is surface plankton feeder. It chiefly feeds on phytoplankton. The gills of fish are specially modified as sieving apparatus for sieving small planktonic food. Adults feed mainly on flagellate, dinoflagellate, myxophyceae and rotifera supplemented by decayed macrovegetation and detritus. This fish also accepts artificial feed such as bean meal, rice bran and flour. It grows very fast, better than catla and attains a maximum size of 1.5 to 2 kg/year and sexual maturity at about 2 years of age. In India its seed is produced by induced breeding and stripping. It breeds only once in a year during monsoon. Fertilized eggs are 3.9 to 4.3 mm in diameter. Its fecundity is 1,50,000 eggs per kg of the body weight.

Cyprinus carpio (Common Carp)

It is characterized by a deep body and short head. Mouth is terminal protractile with smooth and plain lips. Generally two pairs of prominent barbells present of which sometimes one pair is rudimentary. Dorsal fin long, originating from opposite the ventral fin. Lateral line is complete. The three varieties are distinguished as follows:

1. *Cyprinus carpio var. communis* (scale carp) characterized by small scales.
2. *Cyprinus carpio var. specularis* (mirror carp) with large shiny and scattered scales, and
3. *Cyprinus carpio var.nudus* (Leather carp) by the leathery appearance due to absence of scales.

Cyprinus carpio var. communis (Bangkok strain) was introduced in India in 1939 and 1957 from Sri Lanka and Thailand. The scale carp and mirror carp thrive in plains; leather carp is mostly confined to cold upland waters. It is an omnivorous bottom and voracious feeder. Under favourable conditions, it attains the size of 1 to 2 kilogram weight in a year. The fish is prolific breeder, breeds throughout the year. In tropical climate the main breeding season is from January to April. They mature within 6 to 8 months. The male matures earlier than females. The fecundity of fish is 1,20,000 eggs per kg of the body weight. Eggs are the smallest than those of Indian

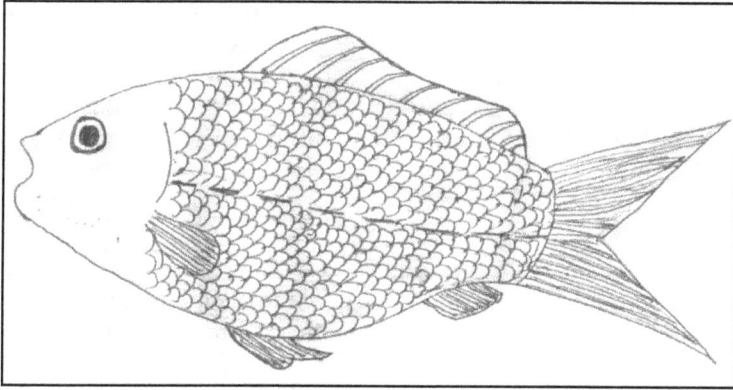

Figure 17.5: *Cyprinus carpio* **var.** *communis*

major carps and other exotic carps. Eggs are yellowish in colour and adhesive, which remain attached with aquatic weed and other artificial substrate provided.

Minor Carps

Labeo calbasu

It is commonly known as 'black rohu' or 'calbasu'. Its body is bluish-green. Small tapering head with subterminal fringe-lipped mouth and four black barbells and dorsal fin with 12 to 13 branched rays are its identifying features. Body is moderately elongated and abdomen is rounded. Eyes are moderately large. Dorsal fin commences slightly in advance of the ventral fin and about midway between the snout and the base of the caudal fin. The caudal fin deeply forked. Ventral fin originates below the fourth dorsal ray and do not extend behind the origin of the anal fin but the anal fin reaches the base of the caudal fin. It is an omnivorous bottom feeder on detritus and animals. It attains a maximum size of one meter and growth in the first year is 30 to 35 cm and 450grams. It is distributed throughout India and found in almost all the rivers, ponds, lakes, tanks and channel.

Labeo gonius

Body elongated and abdomen is rounded. Dorsal profile more convex than that of abdomen. Head is fairly large. Mouth is narrow and inferior. Lips thick and fringed. Both jaws are provided with horny covering on the inner side. One pair of rostral and one pair is of maxillary barbells are present. Dorsal fin commences nearer to the snout than to the base of the caudal fin. Pectoral fin is about equal to the head excluding snout. Caudal fin is deeply forked. Lateral line is complete.

Labeo bata

Body elongated and abdomen is rounded. Lips are thin and continuous, the lower reflected form of the mandible. One pair of maxillary barbells is present. Dorsal fin is with 9 to 10 branched rays. Caudal fin is deeply forked. Lateral line is complete. Scales are large and moderate. Body colour varies with the age of fish, generally silvery, darkest along the back and with the lower fins stained orange. Fine black dots are present on all the fins. It is omnivorous bottom feeder.

Cirrhinus cirrhosa

It has a small head with blunt snout. Lips are thin. Dorsal fin is with 14 to 15 branched rays. The first few rays are much elongated. It has a silvery body and scales with reddish dash except on the abdomen. It is a bottom feeder on detritus and occasionally on zooplankton. It is found in abundance in the river Godavari, the Krishna, and the Cauvery.

Puntius sarana

Body is moderately elongated. Dorsal profile more elevated than ventral profile. Mouth arched and non protrusible. Lips are thin and jaws are simple without any knob or tubercle at the symphysis. Lateral line complete. A pair of rostral and maxillary barbels is present. Rostral pair about as large as the orbit while the maxillary pair is slightly longer. Body colour is silvery. Dorsal side black which gets fainter on the sides and silvery white on the abdomen. Opercle shot with gold and usually with dark spots behind the opercle. There are horizontal pigmented bands along the rows of scales in the upper half of the body. It is an omnivorous feeder on detritus, filamentous algae, microvegetation, worms, insects and gastropods. It is distributed in India, Bangladesh, Thailand, Java, Sri Lanka, China, Nepal, Pakistan, Burma and East Indies.

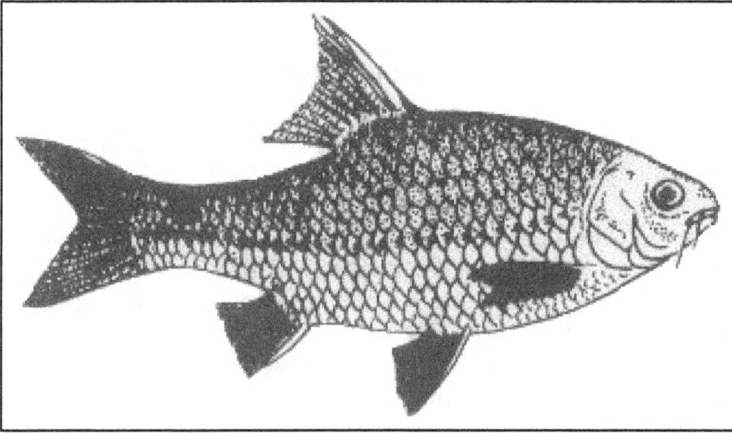

Figure 17.6: *Puntius sarana*

Clarias batrachus

It is popularly known as 'magur' or 'walking catfish'. Magur is widely distributed. It is found across southern Asia including Pakistan, Eastern India, Sri Lanka, Bangladesh, Myanmar, Thailand, Malaysia, Indonesia, Singapore, Borneo, Laos, Philippines, Combodia, Hongkong and China. It is a hardy fish, which can thrive where many other fish struggle to survive. In addition to lakes, reservoirs and lakes, they can be found in brackishwaters or warm, stagnant, often hypoxic waters such as muddy pond, canals, ditches, swamps and flooded prairies.

Body is divided into head, trunk and tail. It is characterized by its spikeless dorsal fin, which extends all along the body; pectoral

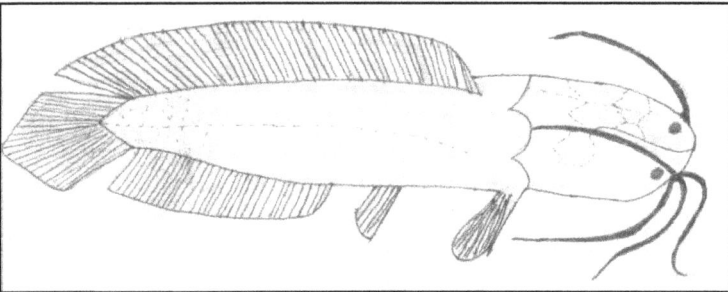

Figure 17.7: *Clarias batrachus*

fin is inserted very low; anal fin is not confluent with caudal. Head is flat with four pairs barbels. The nasal ones reach up to the base of the occipital process and slightly longer than the mental barbels. Maxillary, the longest ones reach the base of pectorals. Mandibular barbels are a little shorter than the maxillary ones. Head bones are superficially exposed. Body is covered by scales and naked skin. Dendrtitic accessory branchial apparatus supplements gill respiration and hence fish can live for a very long period outside water. Eyes reduced and spiracles absent. Parietals, symplectics and sub-operculum absent. Tail is laterally compressed, diphycercal and having rounded caudal fin. The pectoral fins cause painful wounds. They are placed very low along ventro-lateral angles of abdomen. All fins are usually with reddish margins. Body is tinged dark brownish superiorly and becoming lighter beneath. Dorsal, anal and caudal fins are fringed red. Magur can remain dormant through periods of drought and go several months without eating. When they do eat, they consume a wide variety of prey. The fish is voracious, opportunistic feeders who are mainly active at night. They consume a wide variety of prey including eggs and larvae of other fishes, small fishes, a number of invertebrates including crustaceans and insects and sometimes plant materials. In densely populated drying pools, these fish become even more indiscriminate and quickly consume most other species present. The pectoral fins, one on each side, have rigid spine like elements. To move outside of water, the fish uses these spines and flexes its body back and forth to 'walk'. They are well known for their ability to 'walk' on land for long distances, especially during or after rainfall. It is cultured in many countries like Thailand, Indonesia and Philippines, where very high production is reported up to 100 t/ha/year. Its culture is however yet to be picked up in many parts of India. The culture of magur has good prospects for developing domestic trade, because of its high market price, medicinal value, better taste, rich protein content and of less spines. It has present market rate of Rs. 50-75/kg and even fetches Rs.100-150/kg in metropolitans like Hyderabad. It is in great demand in the north-eastern part of India particularly in West Bengal, Assam, Orissa and Bihar for its high nutritional value. It contains higher percentage of protein and iron as compared to other edible species of fish. Its fat content is also very low and is therefore easily digestible so that it is very useful during convalescence. With most other it is a delicacy because of the characteristic aroma and softness of its flesh.

These fishes are not only able to thrive in water containing low dissolved oxygen but also they are extremely hardy with respect to all other environmental parameters and are suited to shallow and derelict waters as well as in saline impoundments having salinity up to 10 ppt. It requires a relatively small area for culture. It can be stocked at higher density than any other culturable species and even in perennial stretch of water as the culture period is restricted to 5-6 months. Fish can live in water without oxygen and in water with a high content of CO_2.

It is carnivorous in habit. They prefer to eat decayed protein food and are considered as scavengers in pond. In cultured operation, they adapt themselves excellently to supplementary feeding with fishmeal, oil cake, rice bran, earthworms, insect larvae, silkworm pupae etc. Adult magur feeds on insect larvae, shrimps, worms, small fish and organic debris found in pond bottom. Fry feed on protozoans, small crustaceans, rotifers and phytoplankton.

The sex of magur can be identified by the pointed anal papilla of the male and oval shaped papilla of the female. Sex identification is possible when the fish attains about 20 cm in length or of one-year age. The anal papilla of the adult male fish during breeding season remains pointed with reddish oval, swells up getting vascularised and the belly becomes distended.

Fecundity of magur is very low. Normally a fecundity of 3000-6000 numbers has been observed from fish ranging between 80 grams and 120 grams. Large sized fish above 130 grams are more fecund having fecundity to the tune of 10,000 numbers or more. In nature the fish has been breed only in rainy season. The fish spawns in horizontal holes. It migrates to adjoining inundated paddy fields and spawn in the grassy bottom nests prepared by them. The eggs are round and yellowish brown in colour. The size of egg ranges from 1.2 to 1.8mm. The eggs are adhesive in nature, cling to the grassy and rocky substrate of the nests/holes. Fish breeds at a temperature ranging from 25 to 30°C. Hatching take about 16-20 hours depending on the ambient temperature gradients.

The male magur guards the nest to prevent devouring of the developing eggs by the predators. Yolk resorption usually takes 2-4 days depending on the water temperature.

Heteropneustes fossilis (Singhi)

It is commonly known as the 'stinging catfish' or 'catfish'. The name 'stinging catfish' is well earned, taken from its ability to inflict severe, painful wounds with dorsal spines and to inject a poison produced by glandular cells in the epidermal tissue covering the spines.

Head flattened, body elongated and laterally compressed. The snout is depressed. Mouth is small and terminal. The eyes are relatively small and lateral in position. The gill openings are wide and the gill membranes are free from the isthmus. There are 7 branchiostegal rays. Dorsal fin short, without spine. Barbels long and four pairs. Accessory respiratory organs are in the form of a pair of lung like sacs, arising from the gill chamber and embedded in the body muscles. The fish is extremely hardy and can able to withstand severe drought conditions with aid of accessory respiratory organs.

It is highly prized air-breathing fish from the Indian subcontinent and Southeast Asian region. The range encompasses India, Thailand, Bangladesh, Pakistan, Sri Lanka, Myanmar, Nepal, Cambodia and Indonesia. Singhi is highly nutritive, recuperative and possessing of medicinal properties. The fish is well known for its invigorating qualities. Its flesh is rich in protein and iron, and fat content is lower in comparison to that in other fishes.

The fish is carnivorous and bottom feeder. Externally sexes can be distinguished only during breeding when secondary sexual characteristics become prominent. The abdominal portion of female is well-rounded. The males look lean. In a mature female, the genital papilla remains in the form of a raised prominent structure, round

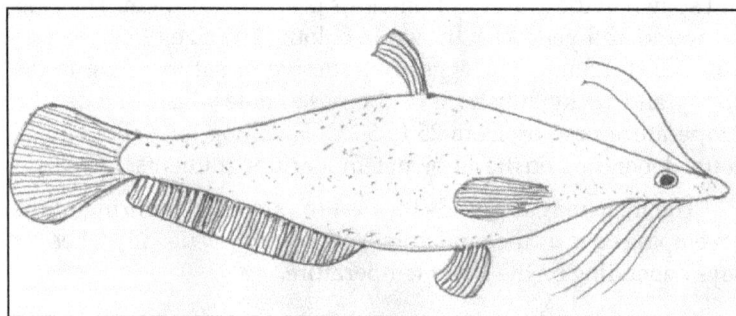

Figure 17.8: *Heteropneustes fossilis*

and blunt with a slit –like opening in the middle. In males, it remains in the form of a pointed structure. Sexual maturity is attained at the end of the first year of life at 8.5 cm. The fish breed in confined waters during the rainy season but can breed in ponds and ditches when sufficient rainwater accumulates. The fecundity varies from 75000 to 1,36,000. The fertilized eggs are adhesive in nature. They are green or brown in colour. The incubation period varies from 18 to 24 hours depending on water temperature. Hatchlings average 2.75 mm in length. Yolk gets completely absorbed by the end of the 4th day, but the larvae commence feeding from the 3rd day. Ciliates and rotifers serve as the choicest food of the larvae at this stage. Aerial respiration starts on 8th day of development. The survival rate is low, which may be due to cannibalism. The fish devours its own eggs and fry when there is lack of food.

Anabus testudineus (climbing perch)

This species naturally occurs in India, Sri Lanka, Pakistan, Burma, Thailand, China and Philippines. Body is oblong and compressed posteriorly. Head is conical. The mouth is broad. Lower jaw is slightly longer. The gill covers are serrated. The fins are well developed. Dorsal fin consists of anterior dorsal fin and posterior dorsal fin. Similarly anal fin is divided into anterior and posterior anal fin. Caudal fin is rounded and numerous fin rays support the fin. The pectoral fins are located near the opercular openings. Colour of the body is light green to dark green above, greenish yellow to

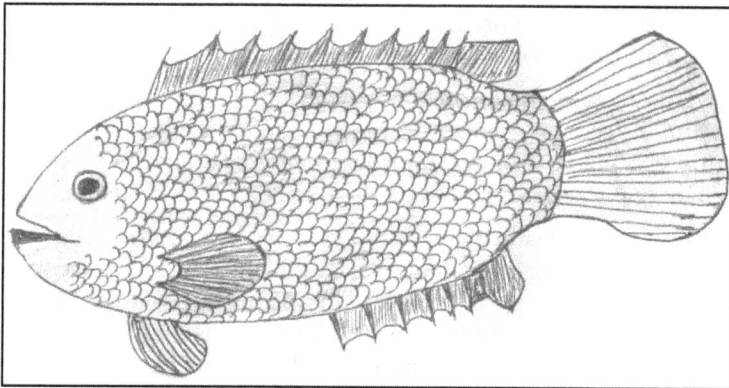

Figure 17.9: *Anabus testudineus*

orange below. Accessory respiratory super-branchial organ is well developed, having thin and folded bony laminae covered with mucous membrane. These breed in confined waters like ponds and lakes. They attain sexual maturity when about 80 mm long and about 6 months old. The breeding season is May to October. The eggs are yellowish or whitish and floating freely on the surface until they hatch. The fecundity varies from 10,000 to 36,000. These attain a length of about 120 mm at the end of first year and about 200 mm at the end of second year.

Channa marulius

It is distributed in India, Bangladesh, Thailand and China. It inhabits lakes, reservoirs, and river but prefers deep clear stretches of water with sandy or rocky bottoms. It is commonly known as giant murrel.

Body is elongated and subcylindrical anteriorly. Head depressed and covered with moderate sized scales. Snout somewhat obtuse. Mouth opening moderate, jaws equal. Teeth on jaws, vomer and palate. Eyes lateral in position and lie in anterior part of head. Accessory respiratory organs in the form of thin bony laminae present in suprabranchial chamber. Dorsal fin long without spine. Both dorsal and anal fins free from caudal fin. Caudal fin rounded. Lateral line complete, which runs straight up to 17th scale and then descends down. Colour varies with age and locality. Greyish green on dorsal side, and pale yellow below 4-5 black blotches located along lateral line. There is a large black spot with orange boundary at the base of caudal fin. It grows up to 120cm in length. It attains maximum size of about 30cm at the end of first year.

Larval stages feeds on zooplankton including copepods and rotifers. It feeds more actively during August to November and January to March. Minimum feeding intensity is observed during spawning period. It is highly predaceous, even cannibalistic.

Fish attains maturity in second year of life. Matured male was reported when size was 300 mm in length and 235 gms in weight. While in female length was 320mm and weight was 252 gms. Fecundity in a length group of 500-820mm ranges from 2214-18475 eggs. The fertilized eggs measure about 1.9mm in diameter and are brownish in colour. It breeds from April to October in natural waters.

Channa punctatus

It is distributed in India, Afghanistan, Pakistan, Nepal, Bangladesh and China. It is commonly found in freshwater ponds and rarely in flowing waters. It is commonly known as spotted murrel.

Body is elongated, subcylindrical anteriorly. Head depressed and covered with large and irregular scales. Snout somewhat obtuse. Lower jaw longer. Teeth present on jaw, vomer and palatine. Accessory respiratory organs in the form of thin bony laminae present in gill chamber. Dorsal fin without spine. Pectoral fins nearer to pelvic fins. Caudal fin rounded. Colour black or dark grey on dorsal side becoming light green or dull on ventral side. Several vertical bands from dorsal side of body pass downwardly to the middle of body. Fins dark grey and spotted. It attains maximum length of 25-30 cm.

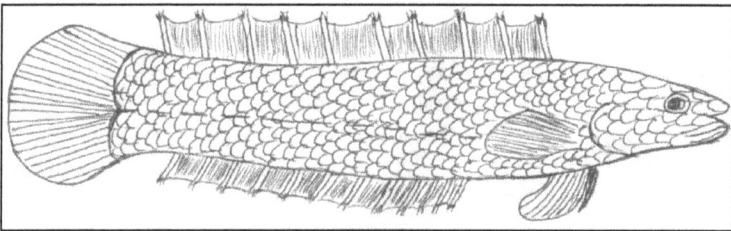

Figure 17.10: *Channa punctatus*

It is carnivorous and predatory fish and feed upon aquatic insects, crustaceans, shrimps and mollusks.

It becomes mature in first year and breeds throughout the year with peak breeding before and during monsoon months. Fecundity ranges from 3000-26000 eggs. It lays eggs in nests in shallow marginal weeds. Parents guard the nest.

Channa striatus

Distributed in India, Bangladesh, Sri Lanka, Vietnam, Cambodia, Indonesia, Thailand, Pakistan, Burma and China. It inhabits freshwater rivers, channels, ponds, lakes, paddy fields, tanks and reservoirs, but prefers stagnant muddy water, swamps and grassy tanks. It is commonly known as stripped murrel.

Body elongated and rounded anteriorly. Head depressed and covered with plate like scales. Mouth opening wide, lower jaw longer. Teeth on lower jaw, vomer and palate. Gill openings wide. Accessory respiratory organs in the form of thin bony laminae present in supra branchial chamber. Dorsal fin without spine. Dorsal and anal fins free from caudal fin. Caudal fin rounded. The colour of fish varies with water. Colour is dark grey or blackish on dorsal side, yellowish white beneath. Grey or black transverse bands on three sides of the abdomen. Ventral and anal fin greyish.

Hatchlings feed on protozoa and unicellular algae and early fry on crustaceans like daphnia and cyclops. Adult feeds on small fishes, aquatic insects and tadpoles.

It attains maximum length of 60-90cm. It may attain a maximum length of 994 mm (3 feet) or slightly more. The growth attained in first and second year varies from 150-160mm and from 310-320 mm respectively.

It attains sexual maturity when two years old. It breeds in two times in a year. Eggs are laid in nests and both parents guard the nest. The eggs are amber coloured, round, non-adhesive and are found floating in a mass in the center of the nest. The average diameter of the egg is about 1.53mm. The fecundity has been observed between 4422 to 20, 000 depending upon the size.

Chapter 18

Induced Breeding

Induced breeding is the technique to stimulate the mature fish to breed in confined waters. Different techniques that are used for induced breeding are described below.

Bundh Breeding

The main problems in riverine spawn collections are dependence of flood positions, admixture of both desired and undesired varieties of fish seed, location and approach to suitable collection grounds, transport facilities etc. To circumvent the problem of availability of quality fish seed, the fishermen of West Bengal established a special type of fishery operation called 'Bundh breeding' in early 1900s in the districts of Midnapur, Bankura and Chittagong (Now in Bangladesh). Bundh is a local term that means a dyke or an embankment. However, with regard to fish breeding, bundhs are special type of perennial or seasonal ponds or impoundments where riverine conditions are stimulated during the monsoon period owing to the accumulation of rainwater from the catchment areas. These are two types, *i.e.,*

1. Seasonal or Dry bundhs.
2. Perennial or Wet bundhs.

Dry Bundh

It is a shallow seasonal pond or a depression on the land, bounded by embankments on three/four sides which impound freshwater from the catchment area during the monsoon season. It remains dry for a considerable part of the year. In West Bengal, a bundh having a catchment area of more than 5 times the area of the bundh proper is considered most suitable. But in Madhya Pradesh a ratio of 1:25 between the area of the bundh and catchment area is considered essential. This ratio depends to a large extent on the average rainfall, in the catchment area. Dry bundhs in Madhya Pradesh are relatively bigger in size (0.2 to 2.5 ha) than those in west Bengal (0.1 to 0.5 ha).

Initially sufficient quantity of rainwater is stored in the dry bundh and brood fishes are introduced into it, from a perennial pond. Spawning is generally observed during and after heavy showers. After breeding is over, the bundhs are partially drained and the eggs are collected for hatchling.

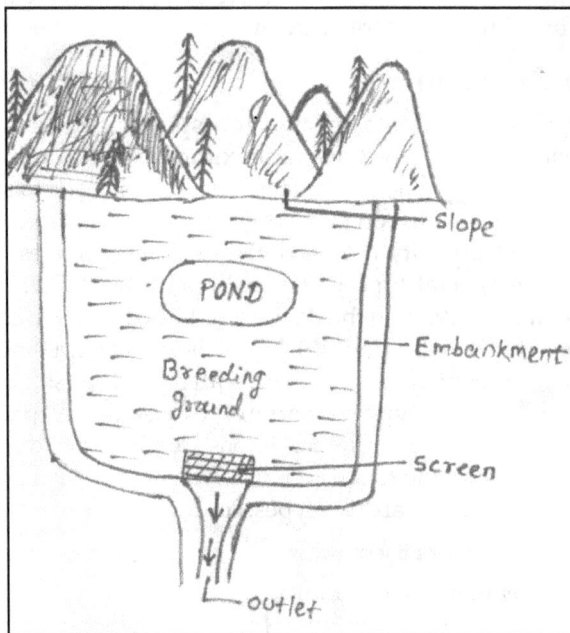

Figure 18.1: Dry bundh

In West Bengal, a seasonal pond is selected in an undulating terrain to receive and store rainwater from the adjoining catchment area and is known as a 'reservoir' or 'storage tank'. Small ponds of 0.1 to 0.5 ha with sloping bottom, having inlet and outlet, are constructed at a lower level than the reservoir. These small ponds or bundhs are connected with the reservoir by channels to have a constant flow of water throughout the period of breeding by gravitational force. The actual number of bundhs depends on the total storage capacity of the reservoir. Both the reservoirs and bundhs are seasonally operated.

Figure 18.2: Modern Mogra Bandh in Maharashtra

In modified type, before the breeding programme the inlet is opened and water is allowed to accumulate up to 30-60 cm. The mature males and gravid females (1:1) are selected and introduced into dry bundh. The brooders are left undisturbed for 6-12 hours to get acclimatized to the new environment. Then intramuscularly injection of fish pituitary extract @ 5-10 mg/kg bodyweight of fishes are administered in a single dose to 10 to 20 per cent of the stocked males as well as females. Females are generally injected with a little higher dose than the males in evening or night hours. Now the inlet and outlets are opened so that rainwater from the reservoir gushes into the bundh and goes out. Proper attention is essential to guard the inlet and outlet with iron meshed screen to check the escape of brooders. The regular water flow is maintained till the sex play starts (sex play starts after 4 to 6 hours of injection). The flow of water is regulated and finally both the inlet and outlet are closed. The breeding continues for 4 to 6 hours. Mostly all the fishes released in the bundh participate in the breeding activity irrespective of whether one received the pituitary extract dose or not. This phenomenon is known as 'Sympathetic breeding'.

After breeding, the eggs are left undisturbed till the embryo is fully formed. Two persons collect the eggs after lowering the water level by dragging a mosquito netting cloth piece. After that, eggs are transferred for hatching to the small earthen pits with mud-plastered walls. These pits are called as 'chaba' and approximately 3'x2'x1' in size. Usually 2 to 3 liters of fertilized eggs (50,000 to 75,000 in numbers) are dumped in each pit and water from the reservoir is circulated through the pits. Hatching of eggs takes place 15-20 hours after fertilization. The hatchlings are collected from the 'chabas' with help of muslin cloth pieces and transferred to similarly prepared bigger sized earthen pits (4'x3'x1½") locally called as 'hamar'. A total number of 3 to 5 lakh hatchlings are usually dumped in each hamar. The selling starts 48 hours after hatchling or even earlier.

In Madhya Pradesh, sufficient rainwater is accumulated in dry bundh before the brooders are introduced. The fish farmers generally prefer a rainy day for this purpose. The hatchling of eggs is processed in double walled cloth hapas or in cement cisterns.

Wet Bundhs

It is perennial pond situated in the slope of a vast catchment area. Wet bundhs are comparatively much bigger in size than the dry bundhs. They have proper embankments equipped with inlets at the up-land side and outlets towards the low-land zone. During heavy rain, a greater portion of the bundh becomes inundated. The excess water flows out through the outlet. A wet bundh can be of any shape and dimension. Normally, the catchment area of 20 to 100 times the area of the bundh is considered better for breeding. With the onset of monsoon, the rainwater from the catchment area

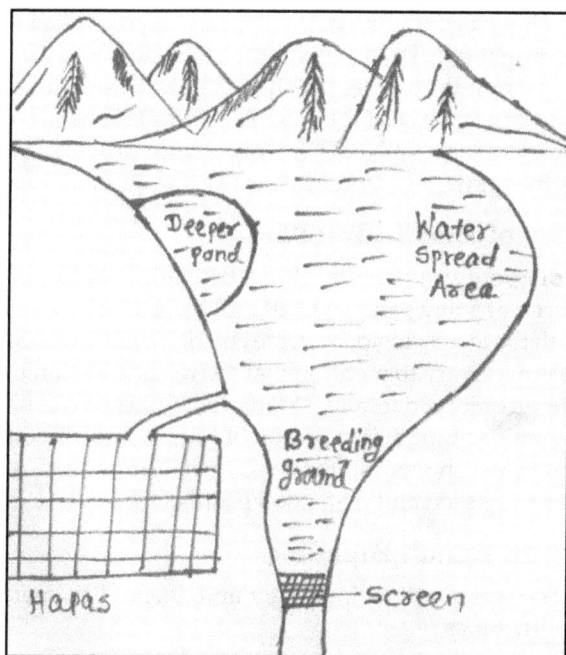

Figure 18.3: Wet Bundh

gushes to the bundh where the brood fishes are already present. In the wet bundh when the conditions are favorable, the fishes start breeding and the eggs are collected and hatched in the same way as in dry bundh. Construction of wet bundh is extremely expensive and not a profitable proposition. However, the existing bundhs can be developed and improved for carp seed production.

Several factors are responsible for spawning of carps in bundhs. Spawning is stimulated by heavy monsoons that flood the shallow spawning areas. It appears that availability of shallow spawning grounds inundated with fresh rainwater is an important factor in stimulating the brooders. According to some workers, spawning occurs in still waters, while others believe that strong current on the spawning grounds favours breeding. The favorable temperature of water for spawning has been found varies from 24 to 32°C under various environmental conditions. Generally the cloudy days followed by thunderstorm and rain, is regarded to influence spawning. Other factors like pH, alkalinity, high oxygen content etc; cannot be ignored. These factors are associated with the floods. Generally, the bundhs having highly turbid waters with a distinct red colour, low pH between 6.2 to 7.6,5 to 8 ppm of dissolved oxygen, low total alkalinity and 27 to 29°C temperature provide favorable conditions for spawning in bundhs.

Economics of Bundh Breeding

In Midnapore and Bankura (West Bengal), 75 lakhs of spawn was produced at a time, and 160-220 million spawn produced in a season. With the increasing pace in the creation of a large number of bundhs, it is necessary to mention that spawn production through dry bundhs, is quite economical. Many crops of seed can be easily obtained from one bundh in a season of four months. By utilizing the rainwater, which would otherwise have been wastewater; we can produce carp seed and reap good profits.

Problems in Bundh Breeding

1. The wet bundh being larger and deeper has operational difficulties
2. Percentage of the collection of eggs is poor due to large size of breeding area
3. Leftover eggs perish due to rapid depletion in oxygen
4. Eggs and hatchlings are preyed upon by the predatory and weed fishes
5. The seed production in wet bundhs is not so pure, because it harbours some weed and predatory fishes which breed along with carps.

Hypophysation

In India, induced breeding of Indian major carps was carried out for the first time in 11957 by Chaudhari and Alikunhi at Central Inland Fisheries Research substation, Cuttack (Orissa). Now a days, in addition to Indian major carps, Chinese carps, catfishes, mullets etc. are induced to breed by hypophysation. Hypophysation technique includes the following steps:

Selection of Donar Fish

The fish from which the pituitary gland is collected is called as donar fish. The donar fish should be fully ripe and gravid. Glands collected from immature or spent fish usually do not give satisfactory results on account of their relative low gonadotrophic potency. The most appropriate tie for gland collection from Indian major carps is just prior to or during the breeding season. It is best to collect the glands from freshly killed fishes but those collected from ice-preserved specimens are also used. Glands can be taken from either male or female. Glands from the same species as the recipient fish (homoplastic) or from the closely related species (heteroplastic) can be used for hypophysation.

Collection and Preservation of Gland

The common adopted technique of gland collection is by chopping off the scalp of the fish skull right above the eye by affecting an oblique stroke of a butcher's knife. After the scalp is removed, the grey matter and fatty substances lying over the brain are cleaned with a piece of cotton. The brain thus exposed, is carefully lifted out by detaching it from the nerves. The membrane covering the gland is then picked up intact.

Pitutary gland is also collected from the posterior hole of the skull called foramen magnum. It is a much easier method. In this method the fish is deheaded. At the back of the head, the foramen magnum can be clearly seen holding the grey matter and fatty substances in it. The brain lies in the ventral side of the foramen. For taking out the gland, the grey matter and fatty substances are first removed by inserting the blunt end of forceps. The brain is then lifted above carefully and pushed forward or pulled out of the hole. The gland lying at the floor of the brain box is then picked up using a pair of fine tweezers. The gland is then preserved in absolute alcohol for defattening or it is immediately frozen. The absolute alcohol preserved gland can be kept at room temperature.

Preparation of Gland Extract

For the preparation of the extract, the required quantity of the gland is taken out from the absolute alcohol and kelp on a piece of filter paper five minutes so as to allow alcohol to evaporate. The gland is then homogenized with a little quantity of distilled water or 0.3 per cent sodium chloride solution (saline water). The homogenized gland is diluted @ 20-40 mg of gland per ml of distilled water. The gland suspension is then centrifuged so as to separate the suspended particles, crushed tissue and fat bodies. Clear fluid containing the hormone is either used immediately or preserved in glycerine (1 part extract and 2 parts glycerine).

Dosages of Pitutary Extract

Females having a round, soft and elastic abdomen with swollen reddish vent and males with freely oozing milt are selected for breeding. A male fish can also be distinguished by feeling the denticulations on the dorsal side of the pectoral fins. The dose of pitutary extract mainly depends upon weight and stage of maturity of the recipient fish and climatic conditions. The potency of the gland also varies according the size and stage of sexual development of donor fish as well as its species, time of collection of gland and its proper preservation. Two injections are given to the female and one to the male brooder. Female brooders of Indian major carps are given a preliminary or first lower dose (2-3 mg/kg body weight) followed by second higher dose (6-8 mg/kg body weight)) after 4-6 hours of first dose. Whereas, male brooders do not usually require any preliminary dose. They are only given one dose @ 2-3 mg/kg body weight at the time of second dose to the female. The exotic carps like grass carp (*Ctenopharyngodon idella*), and silver carp (*Hypophthalmichthys molitrix*) are given slightly higher doses *i.e.*, first dose @ 4-5 mg/kg body weight followed by second dose @ 6-8 mg/kg body weight to females and single dose of 4-6 mg/kg body weight to males.

Injections are generally given intramuscularly between the base of the dorsal fin and the lateral line or between the anal fin and lateral line. While giving injection, the needle is inserted under the scale keeping it first parallel to the body of the fish and then pierced into the muscle at an angle of 45° with the body axis of the fish. The time of injection depends upon the environmental conditions.

Ovaprim is also used in place of pituitary gland. It contains salmon gonadotropin RH and Domperidone, which trigger the ovulation in a single dose. The dose of ovaprim recommended for male fish is 0.10-0.20 ml/kg of body weight and for female it is 0.25 –0.80 ml/kg body weight. This fish spawning hormone is manufactured by Syndel Laboratories Ltd. Canada.

Ovatide is an indigenous spawning hormone manufactured and marketed by Hemmo Pharma, Mumbai. When used; it produces increased number of eggs through complete spawning with high fertilization and hatching percentage. It can be injected simultaneously in male and female broodfish and can be administered in single dose without causing any adverse effect on broodfish. It can be stored at room temperature (20-30°C). It is quite effective even under climatic adversities and is available in the market as 10 ml vial, which costs Rs. 300. The dose of ovatide recommended for female Indian carps varies from 0.20-0.50 ml/kg of body weight and for male it is 0.10 –0.30 ml/kg body weight. Female exotic fishes are injected @ 0.40-0.50 ml/kg, while males are given dose @ 0.20-0.25 ml/kg body weight.

Brood Fish Care

Healthy and disease free brooders of 2-4 years age group and 1-5 kg weight are selected and stocked in a separate brood pond @ 1500-2000 kg/ha in the month of January–February, 5-6 months prior to the breeding season (July-August). They are fed on artificial diet @ 3-4 per cent of their body weight. Grass carp is fed on aquatic weeds such as Hydrilla and Vallisneria once or twice a week at the rate of 25-30 per cent of their body weight.

Selection of Brooders

Success of induced breeding of fish depends largely on proper selection of a suitable brooder. During breeding season, fully mature males and females can be easily distinguished. The fully mature male has rough dorsal surface of pectoral fins and oozes milt freely when the belly is pressed. The mature female has soft, rounded and bulging abdomen with swollen pinkish or light reddish vent. The dorsal surface of pectoral fin is smooth.

Breeding Devices

Breeding Hapa

A breeding hapa is a box shaped rectangular container made up of fine meshed mosquito net. The size of hapa varies from 2mx 1mx 1m to 4mx2mx1m. The choice of the size depends upon the size of the broodfish. It is provided with ropes at its four top and bottom corners. It is fixed in water with the help of bamboo poles in such a way that about 15-20 cm of it remains above the surface of water. The hapa is closed from all sides except a portion at one end of the upper horizontal wall provided with a zip which can be closed or opened as and when required. The injected brooders are introduced through the open end and the hapa is closed properly so that brooders do not jump out. One or more sets of brooders are introduced in hapa depending upon its size. A set usually consists of one female and two males. The weight of two males should be almost equal to that of the female.

Breeding Tank

It is a modification of circular or chinese hatchery. It is a circular concrete tank of 8.0 m diameter, 1.5 m depth and with 50 m^3 water holding capacity. Water is supplied to this tank from overhead water tank. The wall of the tank is provided with diagonally fitted inlet pipes for keeping the water in circulation and simulating riverine conditions, which are conducive for fish spawning. In the center, an outlet pipe of 10 cm diameter is fitted, through which fertilized eggs and water pass into the incubation tanks. Injected brooders (40 kg females and 40 kg males) are released in the breeding tank.

Cement Cistern

Cement cistern having water depth of 35 cm is used for breeding purpose.

Stripping Method

Eggs of carps and few cold water fishes can be artificially fertilized by stripping. In this technique, the eggs are forced out from

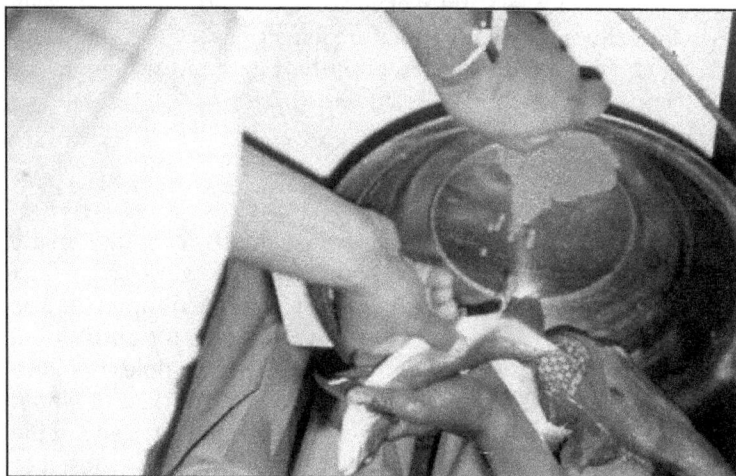

Figure 18.4: Stripping of Female Fish

the body by gently massaging the belly of female fish. The milt is taken from male fish and the eggs are then mixed with milt. Stripping method is of two different types *i.e.*, dry method and wet method.

In dry method eggs and milt taken from stripping are mixed thoroughly and left in this condition for about half an hour. Then water is added in it.

In wet method, eggs are kept in water and milt of the mature male fish is spread directly over into this water. Wet method is commonly used for sticky eggs. Fertilized eggs are transferred to the hatching pits. Generally the stripping is done after the brood fishes are injected with hormone.

Chapter 19

Hatcheries and their Operation

In natural breeding, the hatchlings have to suffer to vagaries of nature and their survival rate is very poor. The fluctuations of the physico-chemical properties of riverine water influence the hatching rate of eggs. High temperature, low pH and dissolved oxygen, high salinity level etc adversely affects the rate of hatchling of eggs. So as to obtain a higher rate of hatching and better survival of spawn, hatcheries have been developed in controlled conditions. In modern hatcheries, the recovery of spawn has increased to 80 to 90 per cent from a lower hatching rate of 30 to 40 per cent, which is generally obtained in traditional methods of hatching.

Factors Affecting the Hatching

Some of the important physico-chemical parameters that affect hatching are described below:

Temperature

Major carps breed during the south-west monsoon period when the temperature ranges around 26-32°C. Optimum range of temperature for better hatching results is 27-28°C. At high temperature, the incubation period is reduced, while in low temperature it is prolonged. Solubility of oxygen is also reduced at high temperature. Mortality of fish eggs occurs at high temperature and the percentage of hatching is reduced.

Dissolved Oxygen

Dissolved oxygen below 6 ppm can be harmful. In low oxygen level, the embryonic development is retarded.

pH

For better hatching percentage, the optimum pH level shall be between 6.4 to 8.5.

Salinity

The optimum range of salinity for better hatching results is 10 to 100 mg/liter

Alkalinity

For better hatching performance, the optimum alkalinity level shall be between 150-250 ppm.

Hatching Techniques

Indian major carps are bred in bundhs or by hypophysation. After breeding the eggs are hatched by different methods as described below:

Traditional Methods

Hatching in Earthen Pits

Hatching carp eggs in earthen pits is a ancient method followed in West Bengal and Bihar. Pits are made of the size 3'x2'x1' and the inner side is plastered with red soil. They are made in rows and are interconnected. About 40,000 to 50,000 eggs are spread in each pit and a mild current of water is maintained with draining facility. The eggs hatch out in pits. After about three days, when the yolk sac is absorbed, the available spawn is collected with a piece of cloth and is transferred to nurseries. This technique being a primitive type, the percentage of spawn realization is subject to cool and favourable climatic conditions.

Hatching Hapa

It consists of two hapas. They are outer hapa and inner hapa. The inner hapa is smaller and fitted inside the outer hapa. They are open and rectangular in shape and are provided with ropes. The inner hapa is 1.5 x 0.75 x 0.5 m. It is made of fine meshed mosquito net cloth. The outer hapa is 2x1x1 m and is made of fine cloth or nylon. These hapas are fixed in a pond with the help of stout bamboo poles in such a way that 25-30 cm hapa remains above the surface of water. The fertilized eggs from the breeding hapa or tank are transferred to inner hatching hapa. About 0.75 to 1.0 lakh eggs can be spread in each hapa. The incubation period for Indian major carps and exotic carps (except common carp) eggs is 14-20 hour at temperature ranging from 24-31°C. After hatching, the hatching wriggle out through the mesh of the inner hapa into the outer hapa. When the hatching is complete, the inner hapa along with the egg cases and unfertilized eggs is carefully removed. Hatchlings in the outer hapa are allowed to remain undisturbed for 2-3 days.

The hapa hatchery has some drawbacks. Due to sudden change in environmental conditions, heavy mortality of fish eggs and spawn occurs. Moreover, this also requires large number of hapas every year, which involve heavy recurring expenditure.

Modern hatcheries: Different types of vertical jars and circular hatcheries have been developed for better hatching of carp eggs under controlled conditions. They are:

1. _Vertical hatcheries_: Glass jar hatchery, transparent polythene sheet hatchery, hanging dipnets hatchery, plastic bin hatchery, modern carp hatchery model CIFE-80, D-81 and D-85.

2. _Vertical hatcheries_: Portable circular hatchery, portable hatchery (Amitava model), cement circular hatchery.

Vertical Hatcheries

Glass Jar Hatchery

This consists of fish breeding unit and a hatching unit in controlled conditions. Fish breeding takes place in the breeding hapas fixed in the cement cisterns inside the hatchery building.

The water is drawn from a freshwater tank and is led to an overhead tank at a height of about four meters. The overhead tank at a height of about four meters. The overhead tank is connected to the hatchery showers are arranged to the breeding tanks and spawn receiving tanks for providing aerated cool water.

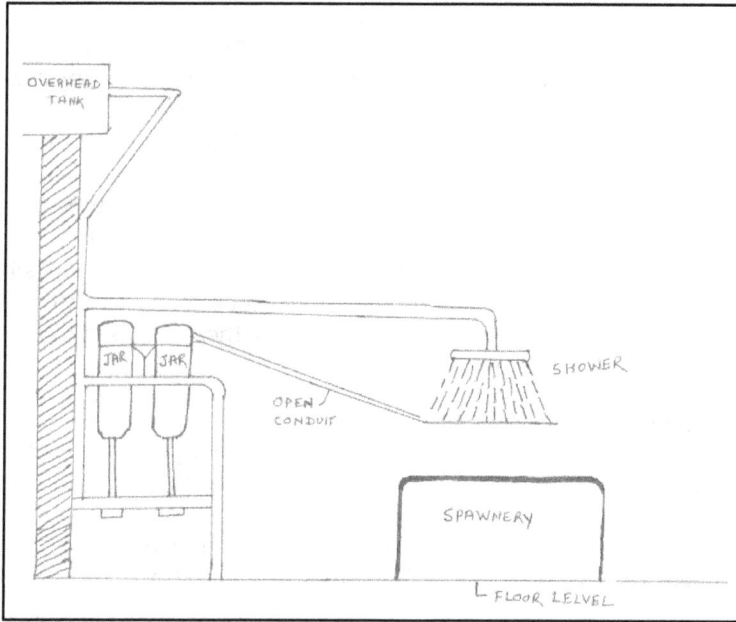

Figure 19.1: Glass Jar Hatchery

The hatching unit consists of a series of cylindrical glass jars that are conical at the bottom and are arranged in vertical rows. Each jar is of 6.35 liters capacity and about 50,000 carp eggs can be hatched. The outflowing water from the spouts of the jars falls into a common conduit, which leads to the spawn-receiving tank. The water enters the glass jar at the bottom and goes out through the spout at the side top, which leads to a common conduit. Midwater flow is arranged in such a manner that the eggs are kept in constant motion and at the same time does not escape through the outflow spout. After hatching is over, the water flow is slightly increased to allow the spawn to escape throughout the spouts into the common conduit and thereby to the spawn-receiving tank. The spawn is retained in the spawnery (spawn receiving tank) for three days until the yolk sac is completely absorbed. Fine cool water shower is provided to the spawn-receiving tank, which replenishes the old water continuously. This gives a better survival rate of the spawn. This system of glass jar hatchery is available at Cuttack in India.

Transparent Polythene Sheet Hatchery

The hatchery is based on the same principle as the glass hatchery. The glass jars, which are costly and breakable, are replaced by a transparent polythene container. The polythene jar is of two-liter capacity having a height of 27 cm and diameter of 10 cm in which accessories similar to these of the glass jar hatchery are used. The jar is provided with an outlet of five cm long polythene tube and inlet pipe. When the eggs are directly taken into the jar, they can escape easily. In order to avoid this problem, an inner mosquito net egg container is used. In each of these inner containers, nearly one-liter eggs are taken and the rate of flow can be then increased to one liter per minute.

Hanging Dipnet Hatchery

This hatchery is installed in CIFA, Kausalayaganga fish farm (Orissa). The water supply comes from an overhead tank fixed at a height of 3.2 meters over the roof. The spawning tank is 2.36 m × 3.23m × 0.9m. There are two breeding tanks of the size 1.2m × 0.7m × 1.06m and two hatchery tanks of the size 3.3m × 1m × 1m.

All the tanks are provided with inlets and drain pipes. Overhead sprays are fixed over all the cisterns and air coolers are fitted for cooling the water. Hatching dip nets are barrel-shaped having steel rings of 65 cm and 46 cm dimension at the top and bottom respectively, covered by 1/16-inch mesh cloth. A 40mm brass spray head is fitted at the bottom of each net.

About one-lakh eggs in each net and 1 to1.5 liter/minute water flow is maintained during hatching. After hatching, the spawn passes into the cisterns and is allowed to remain there for three days until the yolk sac is absorbed, and then the spawn is transferred to nurseries.

Shirgur Linear Bin-Hatchery

There is a common rectangular aluminium tub in which three egg vessels can be kept at a time for hatching more fish eggs. The outer aluminium tub has the size 54″ × 18″ × 22″, which has three chambers. Inlet pipes, outlet pipes and drain pipes are arranged to the rectangular tube to conduct the regular flow of water. The egg vessel is made of 14-gauge aluminium sheet with perforations of 2.5 mm diameter. The egg vessel is cylindrical in shape with 12″ diameter and 12″ height. The plunger lid can be slided and fixed at any given height on a central vertical rod having a series of holes at one cm distance. The plunger lid is used to keep the eggs inside the egg vessel constantly circulating with mild water current and at the same time, it prevents the eggs from escaping from the egg vessel. Each egg vessel can hold about two lakh major carp eggs. Thus six lakh eggs can be hatched at a time from each linear bin hatchery. A commercial model of six compartmental linear aluminum hatchery has also been developed, which works on the above principles.

Modern Carp Hatchery CIFE-D-80 Model

This is a modern carp hatchery where the carps are bred inside the breeding tanks. The eggs are subsequently transferred to hatchery buckets. The process of hatching fish eggs is completely under controlled conditions. The water is led either from an open pond or from a tube-well. If the water is taken from an open pond, it is filtered through filtration chambers to get rid of plankton, silt and other extraneous material, and clear water is led to the over head tank. In case of a tube-well, the water is directly pumped to the overhead tank. The water from the overhead tank passes through the cooling tower where the temperature of water is reduced further. From the cooling tower, the water is connected to the D-80 hatchery bucket at the flat bottom side. The D-80 hatchery bucket is made up of L.D.P.E. material and has a 45-litre capacity. It has a flat bottom. The egg vessel is made of circular aluminum frame to which mosquito netting encasement is provided. About two lakhs eggs can be kept for hatching at a time in the egg vessel. The water is led from the topside of the outer bucket into the spawneries. Water shower is arranged on the top of the spawnery. Artificial aeration is provided by an air compressor or air blower into the egg vessel and subsequently to the spawn-receiving tank. The supply of artificial aeration into the egg vessel keeps the eggs rotating constantly in the water and at the same time supplies additional oxygen to the developing embryos. After hatching, when the spawn is received into the spawneries, supply of compresses air through plastic tubes of three mm and air stones provides additional oxygen for respiration of young hatchlings. Supply of water spray provides well-aerated water. As constant water flow is maintained, metabolities are removed from the spawnery keeping the environment cool and hygienic. The spawnery is a six feet diameter fiberglass or metallic container, about one meter in height. In this spawnery, a clean and smooth hapa is fixed to an M. S. iron the frame. The hatchlings are reared in this container for 3 days until the yolk sac is absorbed. The disadvantage of the D-80 hatchery is that the plastic bucket has a flat bottom and the water inlet is arranged on a side, which does not help the eggs to be kept in a constant rotating condition.

L.D.P.E. Vertical Jar Hatchery (D-81 Model)

LDPE vertical jar hatchery (side elevation) was developed at CIFE in year 1981. The hatchery works on the principles of the D-80 model. But in the D-81 model, the L. D. P. E. bucket has a conical bottom. The inlet pipe enters the bucket from the conical bottom instead of from the side. This helps to lift the eggs vertically and helps them to be in constant rotating condition in the egg vessel.

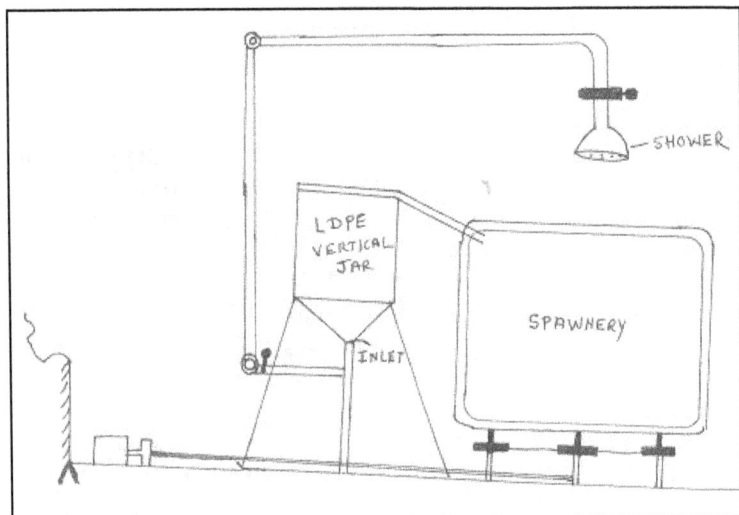

Figure 19.2: D-81 Model

D-85 model

This is a commercial model of the D-81. The material may be L.D.P.E., fiberglass or galvanized iron buckets. In the D-85 model, the bucket is of about 45 liters capacity. The egg vessel is also proportionally bigger. The egg vessel in this case can hold about one million eggs.

If the spawneries are provided with wheels, they can be pushed to a corner after the hatchlings are received and aeration and water spray can be arranged. By providing a new set of spawneries, the D-85 model jars can be used for hatching fish eggs every day.

Circular Hatcheries

Portable Circular Hatchery

This hatchery is made of galvanized iron sheet of 22 gauge with 80 cm diameter and 60 cm height with total water carrying capacity of about 300 litres. The hatchery is exclusively useful as hatching unit and comprises two chambers. The outer chamber is larger one and at the bottom there are eight inlet jets made of copper pipes of 0.8 cm diameter bent at an angle of 600, at equal distances. The water is supplied into this chamber through these jets, which keeps the eggs and hatchings in continuous circulation. At the bottom there is a 3cm outlet, which can be plugged and used as a spawn outlet for harvesting. The inner chamber is28 cm in diameter and is separated by an iron mesh grill which is guarded by monofilament cloth of 40-60 mesh. There is a P. V. C. pipe of 4.0cm and 50 cm height, split in three pieces, and is fixed at the center of the inner chamber. This pipe can be adjusted as per the volume of water in the unit. This pipe serves as an overflow pipe and helps in draining the water when the spawn is harvested. The entire unit is supported on a 2-5 cm iron frame stand. The water supply from the overhead tank to the hatchery unit is fed with a 1.25 cm diameter pipe having a gate valve for regulating the velocity of water flow. The rate of water discharge for a minute is between 8 to 10 litres. About 10 lakh (50 litres) eggs can be hatched at a time in this hatchery.

Portable Hatchery (Amitava Model)

It is prepared from galvanized iron sheet. It is circular in shape. The inside diameter is about 8 feet, and it has a depth of 2 feet 10 inches. The inner chamber of the incubation pool has a diameter of 3 feet 6 inches and is separated by a nylon screen (40 mesh) fitted and stretched on an iron frame. Some 150 liters of eggs (30 lakh) can be kept in the hatchery at a time. The water circulation speed in this hatchery is maintained at 22.5 mt for each revolution.

Cement Circular Hatchery

This hatchery comprises an overhead tank, spawning pool, incubation and hatching tank, and spawn collection tank.

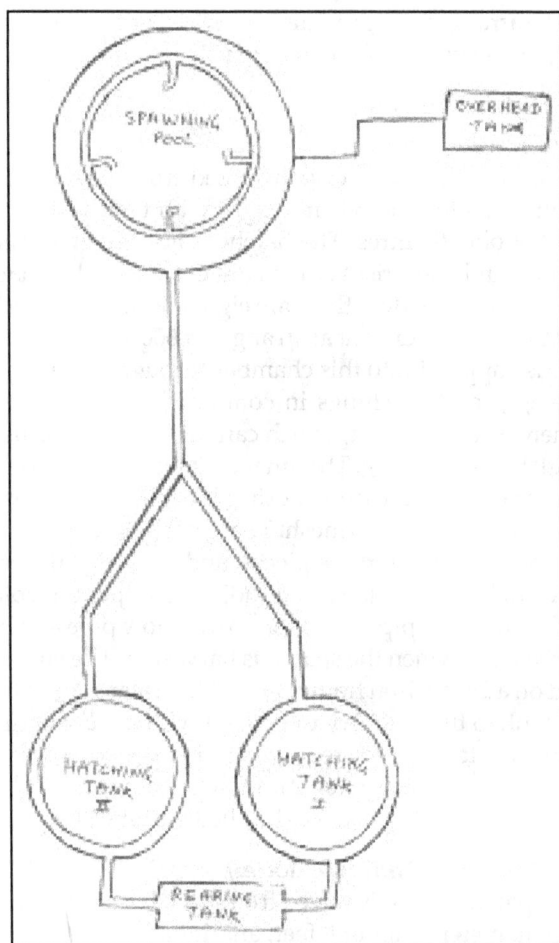

Figure 19.3: Circular Hatchery

Overhead Tank

The overhead tank is made of R.C.C. and has a capacity of about 5000 litres. It is used to supply sufficient water for spawning, incubation and spawn collection tank.

Spawning Pool

It is a circular cement tank of 8-9 meters in diameter and 1 to 1.5 meters depth, with 50 cubic meters of water holding capacity. The bottom of the pool slopes at the center where there is an outlet pipe leading to the incubation tank. The wall of the tank is provided with diagonally fitted pipes for circular water flow due to which riverine conditions are stimulated and the carps breed. About 70 kg of males and 70 kg of females can be used at a time, which will yield about 8-10 million eggs in one operation. After spawning, the fertilized eggs are automatically led to the center and pass through the outlet pipe into the incubation chambers. In order to increase the dissolved oxygen content in the tank, a perforated galvanized iron pipe is fitted above the breeding tank in such a manner that the fine showers fall inside the breeding pool.

Incubation and Hatching Pools

The incubation and hatching pools are circular in shape and constructed of brick and cement. Generally, two incubation tanks are connected to one spawning pool. It has an inside diameter of 3-4 meters and a depth of about a meter with water holding capacity of 9-12 cubic meters. It usually holds about 0.7 to 1.0 million eggs per cubic meter of pool.

Water circulation is very important for proper hatching of fertilized eggs. For this purpose, several diagonally fitted pipes lead to the bottom and sides of the pool. The water inlets at the bottom and sides of the incubating pool are so arranged as to create circular movement of water-the eggs are kept in a constant circular movement. A fine nylon screen separates the inner chambers of the incubation tanks. This nylon screed is stretched and fitted on an iron frame. A rubber belt is fastened very tightly to seal the compartment. In the center is the overflow pipe at a particular level through which the incoming water is led out after a particular height. The water speed is controlled by means of a gate valve.

Breeding Technique

First of all, the breeding tank is filled with water. All the diagonal inlet pipes are kept open to maintain a constant circular motion inside the breeding pool. After injecting, the spawners and milters are released in the breeding pool. Due to the constant circular movement of water, riverine conditions are stimulated. The brood fish get stimulation and breed. The fertilized eggs are led into the incubation chambers through a central outlet pipe, which is controlled with a gate valve. The flow of water in the incubation chambers is controlled to keep the eggs constantly in circular motion due to diagonally arranged pipes. This arrangement helps in proper hatching. The excess water is led out through a central outlet pipe fitted in the center of the inner chamber. The spawn is retained for three days in this, till the yolk sac absorbed.

Spawn Collection Tank

The spawn led into the spawn collection tank, in which a fine nylon hapa is fixed. The spawn get collected in the nylon hapa from where it is counted and transferred to the nurseries.

Chapter 20

Role of Soil and Water Quality in Fish Culture

Water and bottom soil are directly concerned in the culture practices of fish. Physical, chemical and biological properties of soils, denuded and transported from surrounding areas and deposited in the basin shaped areas, vary from those of the cultivated soil from where it has been removed. A number of soil constituents such as colloidal clay, organic matter, nitrogen, phosphorus, potassium etc and also trace elements are carried down with flowing water; get distributed in different properties at various parts of fish ponds which ultimately determines the physico-chemical and biological characteristics of the water of fish pond.

Soil

Physical Properties of Soil

Texture, structure, bulk density and colour are the main physical properties of soil. An ideal pond soil should have sufficient amount (50-60 per cent) of clay and silt in association with other soil particles. Fishponds with gravel and sandy bottom are unsuitable for fish culture. Soil structure with loamy or clay loam bottom and dark in colour is categorized as most productive fishpond.

Chemical Properties of Soil

Soil reaction, organic carbon, available nitrogen and available phosphorus are the main chemical features to categorize the fish pond as low, intermediate or high productive in nature.

pH

Soil reaction is the main factor, as it determines the acidic, neutral and alkaline condition of soil. Most of the essential plant nutrients are released in balanced form under neutral condition of soil pH *i.e.*, between 6.5 to 7.5. Highly acidic soil is unsuitable for fish culture. Under acidic water, loss of appetite takes place among the fishes. Highly alkaline condition of soil creates destruction of soil structure and also retards the fish growth. Microbial activity is also affected by the soil pH.

Organic Carbon

Pond bottom soil with very low content of organic carbon is not suitable for fish culture due to its low productivity in nature. Soil over 1 per cent organic carbon is ideal for fish production. Very high organic content is undesirable because it will result in high C: N ratio. Under high C: N ratio *i.e.*, over 10:1 will adversely affect the microbial activity and most of the soil nitrogen will be utilized by micro-organisms and soil will lose its nitrogen supplying capacity.

Nitrogen

Soils may be having rich source of nitrogen. But depending on the other limiting factors, if release of available nitrogen is restricted, the soil is rendered unsuitable for establishing fishpond on such soil base. Soils with available nitrogen above 50 mg N/100 g soil is considered to be suitable for fish production. Availability of soil nitrogen is mainly affected by microbial population of the soil in association with other environmental factors.

Phosphorus

It is present in various forms in the soil. Important forms are organic (proteinous and nonproteinous) and inorganic (calcium phosphate, iron, aluminuim and pyrophosphates). Availability and release of soil phosphorus is mainly pH dependent. Under alkaline condition, the available forms of soil phosphorus are converted into octocalcium form and hydroxy apatite is formed and under highly acidic condition, insoluble form of iron and aluminium phsophate is formed and phosphorus becomes unavailable to the organisms. Available phosphorus above 10 mg P/100 g soil is favorable for fish culture.

Biological Properties of Soil

Microorganisms play an important role in determining the physico-chemical characteristics of soil. Soil microbes are highly sensitive to the soil reaction. Soil pH at and around neutral is favourable for healthy growth of soil micro-organisms and microbial activity. Soil fertility based on microbial activity was studied in bottom, side and adjoining areas of small reservoirs. Atmosphere is rich source of nitrogen (more than 79 per cent) but it cannot be utilized by land crops, until and unless it is somehow trapped by the microbial activity in the soil. Several forms of microorganisms are present in soil and are of great importance as they help in the decomposition of organic wastes and release of nutrients. Nitrogen cycle is the best example for bacterial activity in the soil.

Water

Physical Properties of Water

Temperature, colour, transparency, turbidity are among the important physical features of pond water.

Temperature

Temperature, very low or very high are not suitable for fish culture though fish can withstand a wide range of water temperature. Water temperature between 15 and 35°C is favorable for healthy growth of fishes. However, for breeding purposes, temperature between 20 and 30°C is ideal.

Colour

Colour of water should be indicative of plankton bloom, throughout the culture period.

Transparency and Turbidity

The transparency of water *i.e.,* the least level of turbidity in water enables the penetration of light to the depths where it is ultimately absorbed. In deep water or reservoir water, the zone up to which the light rays penetrate is called as 'photic zone' and below this, there is a complete darkness and the organisms that require light (*e. g.* algae) cannot live. Sometimes manifestation of phytoplankton due to the application of organic manures in excess adversely affects the productivity of pond. Photosynthetic activities of plants in turbid water get reduced and the fish fauna needing clear oxygen-rich water are replaced by predominantly labyrinthiform fauna in such a situation. Behavioral changes in fish have been noted at clay turbidities above 20,000 mg/l and appreciable mortality has been noted above 1,75,000 mg/l. Clear ponds having clay turbidity within 25 mg/l have been seen to be more productive than those having turbidities above 25 mg/l. Thus, productivity and photosynthesis seem to be inversely proportional to turbidity. Accumulation of humic substances is also harmful for fishery because of restricted light penetration and development of acidity in such waters. Plankton turbidity is generally good for fish production but this also becomes a menace during eutrophication caused by a high planktonic bloom.

pH

The pH is affected not only by the reaction of CO_2 but also by the organic and inorganic solutes present in water. Any alteration in water pH is accompanied by the changes in other physio-chemical parameters. The pH of a water media often gives indication of fertility or potential productivity in a pond. The diurnal fluctuation of pH of a water body should remain in the range of 6.4 to 8.5 in order to support the optimum fish growth. If waters are more acidic than pH 6.5 or more alkaline than pH 9-9.5 for long periods, reproduction and growth have been found to be affected (Swingle, 1961). The pH range of 6 to 9 is most suitable for pond fish culture. While pH more than 9 is unsuitable (Swingle, 1967). At pH 4 or 11, fish will die out with the development of ulcers as a result of severe corrosion of fish body (observation of R. K. Das at Domjur, Howrah, West Bengal, 1993). Therefore, before we go out for pisciculture in any waterbody, we must know its diurnal fluctuation of water pH. According to Jhingran (1988) surface water generally develops higher pH due to phtotsynthesis while water below the zone of effective light penetration showed low pH values.

Dissolved Oxygen

Dissolved oxygen is a vital parameter as it is directly involved in the metabolic activity of flora and fauna of fishpond. The main sources of oxygen in the pond water are (*i*) diffusion of atmospheric oxygen and (*ii*) photosynthetic production by phytoplankton. During the process of photosynthesis, carbon dioxide is reduced to carbohydrate by the chlorophyll bearing organisms and oxygen is liberated. Solar energy is used and stored in the form of chemical energy. Oxygen thus produced during the process of photosynthesis is utilized by other organisms and equilibrium is maintained.

The dissolved oxygen content of warm water fish habitats should not be less than 5 mg/l, during at least 16 hours of any 24-hour period. It may be less than 5 mg/l for a period of not exceeding 8 hours within any 24-hour period and at no time shall the dissolved oxygen content be less than 3 mg/l (Boyd, 1982).

Continued exposure to low dissolved oxygen is considered as a precursor to bacterial infection in fish. Exposure of fish to dissolved oxygen less than 1mg/l is lethal when it lasts longer than a few hours (Swingle, 1969), because the minimum concentration of dissolved oxygen tolerated by fish is a function of the exposure time. Further, the minimum tolerable concentration of dissolved oxygen will vary with species, size, physiological condition, concentration of solutes, temperature and other factors. It has been found that fish can survive at dissolved oxygen in the range of 1 to 5 mg, but its growth and reproduction are retarded.

Carbon Dioxide

Natural waters generally accumulate carbon dioxide through respiration of aquatic plants and animals. In low concentration it indicates productiveness of a water body (15mg/l) but above 15 mg/l, it hinders the uptake of oxygen by fish. Accumulation of dissolved carbon dioxide also increases with dying of phytoplankton and with cloudy weather when photosynthesis becomes limited. At higher concentration of CO_2, pH decreases and water becomes acidic both of which are not conductive for fish health and growth. At night, with decrease of dissolved oxygen, concentration of dissolved carbon dioxide increases and this makes the loading of oxygen more difficult for fish (Boyd, 1979 a). Thus, at lower concentration CO_2 is good for photosynthesis and at higher concentration it is harmful for fish growth and production.

Alkalinity

Alkalinities refers to the total concentration of bases in water expressed in mg/l of calcium carbonate. In most waters bicarbonates,

carbonates or both are predominant bases. Waters having alkalinities below 20mg/l, show sharp rise of fall of pH with small change in the concentration of bases or acids because of poor buffering capacity of such waters. Waters with such poor buffering capacities often becomes unconducive for fish production. Waters having 40 mg/l or more total alkalinities are considered to be more productive than waters of lower alkalinities. The greater productivity of waters at higher alkalinities does not result directly from alkalinity but rather from the greater availabilities of phosphorus and other nutrients at higher alkalinities. It has also been observed that the outbreak of Epizootic Ulcerative Syndrome (EUS) was more severe with waters having alkalinities below 20mg/l (Das and Das, 1995).

Hardness

Traditionally hardness is a measure of the capacity of water to react with soap. It is often described as carbonate (temporary) and non carbonate (permanent) types of hardness. Fishes have been found to be susceptible to diseases in water with hardness below 20mg/l. Productive waters should have hardness above 20mg/l, calcium above 5 mg/l and magnesium above 2mg/l. Very hard water (>300 mg/l) also becomes uncongenial for fish production because of higher pH. Optimum hardness for fish culture has been observed to be around 75-150 mg/l (Das, 1996). Epizootic Ulcerative Syndrome outbreak has been observed to be more frequent in waters with low hardness (Das and Das, 1995). According to Jhingran (1988) and Sugunan (1990) the hardness above 70 ppm is a indicator of the better productivity. Sugunan and Yadava (1992) mentioned total hardness of Hirakud reservoir at lentic and bay sector at 60 and 48 mg/l respectively.

Total Dissolved Solids

Total dissolved solids gives an indication about total concentration of constituents in water. TDS values affect taste of water and water from different areas are characterized as tasteless, sweet, brackish, saline, bitter etc.

Nitrogen

It is one of the basic elements of fertility in ponds and is present in water in an amount of 65 per cent of all dissolved gases at equilibrium. Certain quantities of nitrogen may reach soil and water from the atmospheric nitrogen by rain or incidental lightening. There is also role of soil bacteria that can fix atmospheric nitrogen and, thus, make it available to the green plants. Whatsoever, are the alternative possibilities, the ponds have the major supply of nitrogen from the decaying organic matters *i.e.*, from the putrefaction cycle in which nitrates are the end products. From the productivity of ponds, sufficient quantities of nitrogen compounds, such as nitrates are essentialities when water contains 4 ppm of nitrogen with 1ppm of phosphorus and 1 ppm of potassium the production of plankton is high.

Hydrogen Sulphide

The deeper layer of many ponds, tanks, lakes, and reservoirs may contain significant amounts of the toxic gas, hydrogen sulphides, which is formed by decaying organic matters. In this decaying action, anaerobic bacteria play the main role. Sometimes the gas accumulates in the fishponds having thick layer of organic deposits at the bottom. This gas can be a severe limiting factor in fishponds, as a concentration of 6 ppm of H_2S will yield drastic effects for the aquatic biota in general and can kill common carp in few hours. In such a situation, aeration of ponds by mechanical agitation can help a lot to free the pond of hydrogen sulphide. Unionised hydrogen sulphide is toxic to fish but its ions are not appreciably toxic. Adelman and Smith (1970) have shown that egg survival and fry development in northern pike (*Esok lucius*) are poor due to the presence of 0.06 mg/l of unionized hydrogen sulphide.

Phosphorus

Phosphates and nitrates are essential for fish production in ponds. In all natural waters, these nutrients are found in varying quantities in dissolved form in the bottom soil. If nutrients are deficient, production and growth of organisms in the pond water are adversely affected. The phosphorus is released from the dead organisms in the bottom of the pond of giving rise to the growth of 1ppm of phosphorous. This level of phosphorus is considered as optimum for sufficient growth of plankton.

Potassium

This is present in a cultivable pond to an extent of 1ppm along with 1ppm phosphorus and 4ppm nitrogen. This concentration of the three constituents determines the fertility of the pond and is considered to be most ideal for the growth of planktons. Potassium is released from the bottom soil which is rich in deposits of various kinds. It is released from there and is specially effective in stimulating the growth of aquatic vegetation.

References

Das, Manas Kr. and Das, R. K. 1995. Fish diseases in India-A Review. *Environment and Ecology*: 533-541.

Das, R. K. 1996. Monitoring of water quality, its importance in disease control. Paper presented in National Workshop on Fish and Prawn Disease, Epizootics and Quarantine adoption in India. October 9, 1996, CICFRI, pp 51-55.

Jhingran, Arun G. 1988. Reservoir Fisheries in India. *Jr. of the Indian Fisheries Association*. 18: 261-273.

Sugunan, V. V. 1990. Reservoir Fisheries Management. In: Sugunan, V. V. and Bhowmick, U. (eds.), Technologies for inland fisheries development. Central Inland Capture Fisheries Research Institute, Barrackpore, India, pp 153-164.

Sugunan, V. V. and Yadava, Y. S. 1992. Hirakud reservoir-strategies for fisheries development. Bulletin 66, CICFRI, Barrackpore, India.

Swingle, H. S. 1967. Standardization of chemical analysis of water and ponds muds. FAO Fish. Rep. 44: 397-342.

Chapter 21

Gears and Crafts Used in Fisheries

For catching fish, conventional and unconventional fishing methods are used. Conventional methods involve the use of gears of catching fish, while unconventional methods involve the use of devices that are helpful in creating conditions for easier catching and gears for catching fish.

Gears are the instruments used for catching fish and the crafts provide platform for the fishing operations and carrying crew and fishing gear. The fishing gears along with the vessel, auxillary equipment and men constitute a fishing unit. The size of the unit depends upon the distance of fishing ground from the shore, handling and disposal of catch as well as geographical factors. The amount of the catch depends upon the efficiency of the unit and productivity of the fishing grounds.

Conventional Methods

Different types of gears are used for catching fish. These are grouped into four categories upon their form, function and mode of operation.

1. Impalling gears
2. Hook and line gears

3. Entrapping gears

4. Netting gears

Impalling Gears

These gears are used to catch large sized fishes, which are concentrated in a small area and are visible. Fishes are caught by wounding them with weapons like spear and harpoon. A spear has a long handle (1.5 to 2.0 m length) fitted with metal head with one to several points. A person standing on a boat throws it at a fish. It is also used to spear a fish through holes in the ice or while the fish rest in streams. Harpoon is a spear modified for throwing. It has barbed iron points or blades attached to a 3 m long shaft.

Hook and Line Gears

It is an ordinary fishing device and is popular with fishermen who cannot afford to have other nets. From a single hook, tied to rod, lines with 100-1000 hooks are in use. One end of the line is fixed to some bamboo or anchored boat and the other end is left across the river with a float tied with a heavy sinker through a long rope. The hooks are attached with a small string to the line at regular intervals. Snails, earthworms, cockroaches, frogs, prawn, cater pillars or small fishes are used as bait for catching different species of fish. The fisherman inspects the line frequently and the hooked fishes are taken out and new bait is refixed.

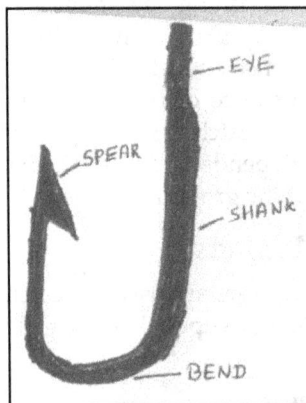

Figure 21.1: A Typical Hook

Figure 21.2: Types of Hooks

Entrapping Gears or Traps

These are the stationary gears and include variety of traps such as basket trap and pot trap.

A basket trap net consists of two dome shaped hemispherical baskets, each provided with an opening at the narrow end, as shown in figure. The opening is guarded by flexible recurved bamboo sticks with their free ends facing towards the inner side. Suitable bait in the form of balls is generally placed in the trap, which is lowered in water for sometime. Fish that enter the basket are unable to get out due to the recurved nature of the sticks guarding the opening.

Sometimes, a wide mouthed earthen pot or vessel is used as a trap. The mouth is closed with a thick cloth having a few holes to provide entrance. Suitable bait placed inside the pot induces the fish to enter the pot, which is placed, on the bottom of the river or pond. Covered pot or plunge basket is used to catch fishes like *channa*, *heterpneustes* and *clarias*. This basket is conical with an opening at the top and is made of bamboo strips. It is about one meter high. It is placed at the bottom of shallow muddy water. The fishermen catch the fishes with hand, through the opening of the basket.

A few species of fish have developed the habit of jumping out of water at intervals or when excited or in danger. Taking advantage of this habit, a horizontal net fixed in water or in a moving boat, is used to catch them when they fall back.

Netting Gears

Nets are the most widely used gears and are of different types. Netting gears catches the major share of world catches.

Net is a piece of webbing in which twines are intersected into regular meshes. Twines are made of natural or synthetic material. Mesh size (distance between two intersections) is very important in selective fishing. Nair *et al.* (1969) found that net of 53 mm, mesh bar were suitable for the capture of *Labeo calbasu* in Gobindsagar reservoir. Kartha and Rao (1991) recommended 148 mm, 89 mm and 60 mm for catla, mrigal, and rohu respectively. 55 mm bar net was found ideal for harvesting *L. diplostoma* and *L. bata* (George *et al.*, 1975). Natarajan (1976) determined the mesh bar 91, 41, 52 mm respectively for *C. catla, C. mrigala* and *L. calbasu*. In Hirakud reservoir, the mesh size for catla was fixed at 90 mm bar (George *et al.*, 1979). A mesh bar between 100-150 mm for catla in Gobindsagar reservoir is recommended by Anon (1980).

Net has accessories like ropes and cables, floats, sinkers or weights, stakes, anchors etc. Floats are attached to the head rope to keep the net in desired position and shape. They are made of cork, wood, plastic, rubber or hollow metal. Sinkers are attached to the foot rope to keep the net in vertical position and are made of lead or iron. Stones can also be used as sinkers. Anchors are made of suitable heavy material. When net is anchored or set in mid waters or at bottom, a marking buoy consisting of small boat with pole and flag is used to indicate the position of the net. Efficiency of the net depends upon the material of twine, size of mesh, floats and buoyancy, sinkers, type of the fish to be caught and the condition of the fishing ground.

Net setting is done in two ways *i.e.*, active netting and passive netting. Active netting means an operation in which the net is moved after setting for fish capturing while in passive netting the net is not moved, once it is set for fish capturing. The net is either set at the bottom with the help of anchors or stakes or it is suspended at intermediate depths with the help of drop lines from larger buoys at the surface. Sometimes it is held suspended near surface by its own float line but the net is attached by means by ropes to larger sinkers at the bottom.

Following types of nets are used for fishing:

1. Impounding nets (Trap nets)
2. Entangling nets (Gill-nets)
3. Encircling nets (Seines)
4. Dragging nets (Trawls)

5. Dip nets
6. Cast nets
7. Blanket nets

Impounding Nets

These are the stationary nets and their principle of action is impounding the fish by leading them into a enclosure through guarded entrance from where they cannot escape. These nets include fyke or hoop nets and pound nets.

Fyke or Hoop Nets

It is used in rivers and flounders in shore water to catch the catfishes. Fyke net is a long cylindrical bag of net supported on a rigid framework called hoop. The hoop prevents the net from collapsing and also provides attachment for the base of net funnels. The net funnels prevent the fish from escaping. The fishes are directed towards the blind end of the net. On each side of the mouth of bag, a wing of net is attached. The net is fixed in shallow water with its wings adjusted obliquely. As fishes strike the wings, they set deflected towards the mouth of the net. The catch is collected from the last pocket. Fyke nets used in shore fisheries are slightly modified. They have long wings with floats on the top and sinkers at the bottom. In shore water the net is with anchors.

Pound Net

This net is normally used to catch migratory fishes. The enclosure or pound of net is made in various designs. The net may be set on bottom with stakes or held floating with the help of anchors. The lead (wall of net) directs the fish into main chamber or heart, whose gate is guarded by slot like entrance. The heart is connected to several pockets called cribs. The length of lead ranges from 300-800 m. This net is operated in water depth ranging from 2 to 40 m.

Entangling Nets

These are the stationary nets and their principle of action is that while trying to swim through the net, the fishes get entangled in it. There are two main types of entangling nets *i.e.,* the gill net and the trammel net.

Gill Net

Gill nets are wall like nets with floats attached to the head rope and sinkers fixed to the footrope. They are made of cotton or hemp of

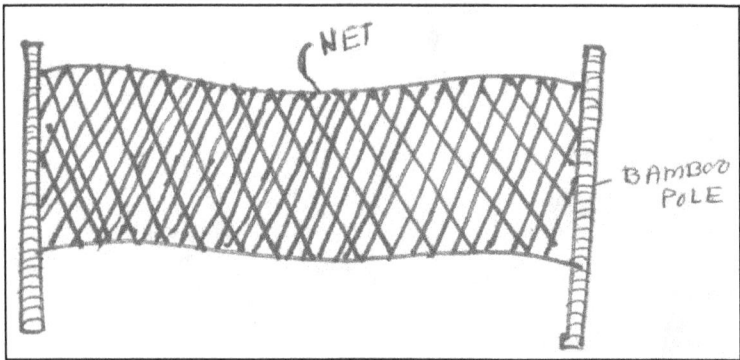

Figure 21.3: Stationary Gill Net

different sizes of mesh. The net is set in transverse direction of the migrating fish, so that when the fish tries to swim through a net wall, the meshes form a noose round its head and the fish is caught. As the fish tries to escape, it gets stuck up behind the opercle, hence these nets are known as 'gill-nets'. In order to be efficient, the meshes must of right size and shape for the fishes to be caught. An important feature of the gill net is that it should contrast as little as possible with its surroundings, so that it does not look like an impenetrable wall, which the fish could avoid. Hence various dyes are used to make the net invisible.

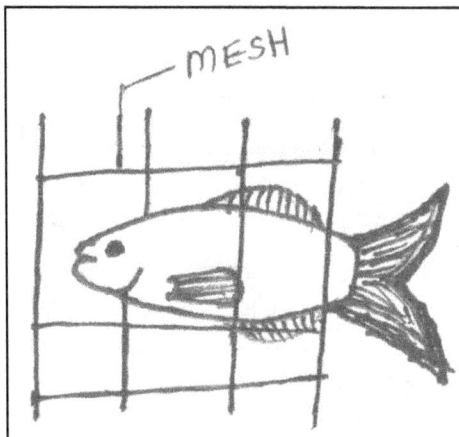

Figure 21.4: Fish Caught in Gill Net

During the recent years, the efficiency of the gill net has been increased considerably by using transparent synthetic fibers instead of natural fiber of cotton or hemp. These nets are generally used to catch big varieties of fish and are therefore made of strong material with large sized mesh. In the evening, the nets are stretched across the riverbanks and fixed by poles. The net is hauled up in the morning and the fish entangled are collected. When the current of water is strong several nets are fixed one behind the other.

Two types of gill nets are used in fisheries *i.e.,* set nets (stationary nets) and floating nets. Stationary set gill nets are net walls set on the bottom between anchors. They are used in large lakes or coastal fisheries and are set as straight walls or in a bow-shaped pattern. Several nets may be combined to form long blocking walls, and long poles are used an anchors. Floating gill nets are not firmly fixed to poles but are simply anchored on the bottom and are suspended by floats on the surface of water. They are called 'floating nets'. They can be set at any desired depth and are made of finest and scarcely visible material.

Tramel Net

It consists of three vertical walls of webbing united at the top to a float line and at the bottom to a lead line. Middle wall of net having small mesh size and made of light twine is hung loosely between the two outer walls which have large mesh size and are made of heavy twine. Small sized fishes get gilled in the middle wall. A large sized fish pass through mesh of first wall and pushes a bag of loose net, through large meshes of third wall. The fishes get entangled and cannot escape. As compared to gill net, the trammel net catches fishes of different sizes. Its mode of operation is similar to that of gill net.

Encircling Nets

Encircling nets are also known as seines. They are of active fishing type. The principle of these nets is to surround a certain section of water mass with a net and to block the escape of fishes. Then by hauling the fishes are led into bag of the net. The efficiency of these nets depends upon various factors such as 1) area of the water mass to be fishes 2) swimming speed of the fish, 3) speed of the seining operation, 4) depth of water, 5) water current, and 6) strength of the net.

Encircling nets of different design, length, width and weight are used in lakes and rivers. Nets should be very strong and sturdy enough to withstand pulling force of hauling, water resistance to net during hauling and the weight of the captured fish. Three types of seines are commonly used. These are haul seine or beach seines, drag seine or Danish seine and purse seine.

Haul Seine or Beach Seine

It is a strong strip of strong netting, hung to a stout float line at the top and a strong lead line on the bottom. The wings of the net are of larger mesh than the middle portion or bunt. The wings are tapering towards ends. The central part of the bunt is sometimes transformed into a bag for confining the fishes. At the end of each wing, the float line and lead line are tied to a stout pole (spreader). The hauling lines are also tied to the pole. At regular intervals, there are spreaders running from float line to lead line so that net hangs properly. The net is operated from the beach and landing also takes place at the beach. For operating the net, its one end is held at the bank of the river while a boat carries the other end to spread it out in the water, in a semicircular way and brought it back to the bank. For hauling, the two ends of the net are slowly pulled by fishermen. The fishes are collected from the bag.

Drag Seine or Danish Seine

It may be without bag. Generally it is smaller than the haul seine. Its wings are smaller, mostly replaced by warps. It is operated in similar way as the haul seine. It is used for fishing in deep water.

Purse Seine

It is used to catch pelagic fishes. It consists of a single strip of net with same width throughout. There is no bag, but the central part of the net has thick twines. The net has a float line at the surface and the lead line at the bottom. On the lower side, net has purse line of strong rope. The purse line is for pursing the net. The net is operated from large vessel (boat) in open water and landing also takes place on the vessel. One end of the net is held on a small boat while the other end is carried by a large boat to spread the net in a circular way on water surface and bring it back to the boat. The hanging net encircles a shoal of fishes. Then the purse line is pulled and net takes the shape of the purse. Finally the entire net is hauled along with the catch.

Dragging Nets

These are of active fishing type and function on the principle of catching fish by moving a net (towed to boats) through water. Varieties of fishes living in mid-water and near bottom are caught. Commonly used dragging nets are known as trawls.

The trawl is a bag shaped net with a mouth and a tapering blind cod end. Upper margin of the mouth is lined with float line and lower margin with foot rope. Wooden or metallic bobbins are attached to foot rope. The cod end is provided with toothed chafing gear and cowhide chafer. Bag is supported by strips to prevent it from collapsing. The mouth of the net extends out into wings. The mesh size decreased from wings towards the cod end of the net. To provide strength to net for trawling, additional lines are attached towing along with the main lines. The length of the net webbing is 1-1½ times that of the float line. The lines are attached to spreader of both sides. The trawls provided with otter board are known as otter trawl. The otter board acts like under water kite. The boards, one on each side, keep the wings apart and mouth of the net open during the entire operation. The trawl in which wings are kept apart by a beam is called beam trawl. Heavy metal frame attached to lower side of beam keep the net one-meter above the bottom. The trawls are either dragged by one or two vessels.

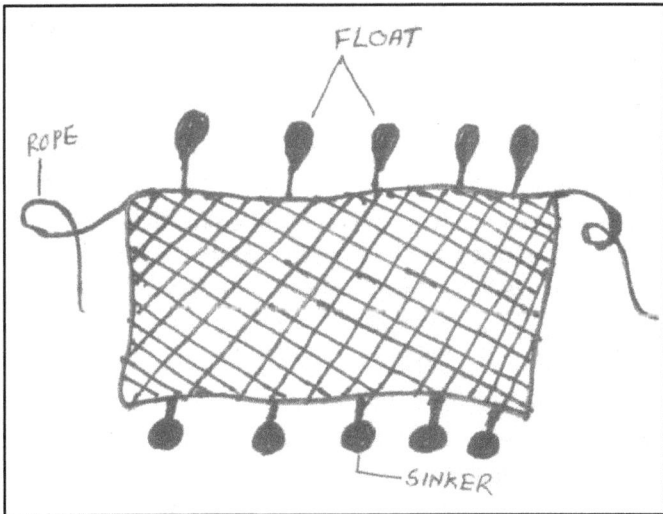

Figure 21.5: A Drag Net

Dip Nets

Shape of dip nets may vary from region to region. These nets are lowered into the water in the hope that fish would swim over them and lifted out of water. They are triangular, rectangular or square in shape and are fitted round a frame. Some bait is often put on the net or suspended over it to attract fishes. The smaller nets are operated by hand, while the larger ones are dipped and lifted from the water by means of a long pole, which is operated like a lever. The original simple hand dip nets have gradually been transformed into permanent stationary lift nets, operated from bridges made of bamboos. The different types of lift nets are described below:

Hela Jal

It is a triangular net. It is made of cotton and consists of two light bamboo sticks crossing each other near one end, supporting the net. A short cross stick is fixed near the apex of the triangle and a bamboo pole works as a handle. To operate, the broad end of the net is pushed along the bottom and then lifted, with a jerk to throw the fish towards the handle. This net is used to catch catfishes and prawns.

Kharra Jal

It is constructed similar to that of hela jal and is a large sized triangular dip net. It is operated from a platform made of bamboo pieces and implanted in the river bed. Fishermen using a bamboo pole as a lever to lower or lift the large nets operate the net.

Bhesal Jal

This is also similar to the above but is operated from a boat.

Khoursla Jal

It is rectangular in shape. It is suspended from the ends of two curved bamboos crossing each other at right angles. The net is lowered in shallow water near the bank of the river by means of a strong rope and is lifted suddenly when fish pass over it. It is generally used for catching freshwater mullets.

Cast Nets

The most common cast net used in rivers and ponds of Uttar Pradesh, West Bengal, Bihar, and Orissa consists of conical netting and a hand rope. The netting is made of nylon twine (formerly of cotton), which is skillfully woven to form meshwork. The mesh size

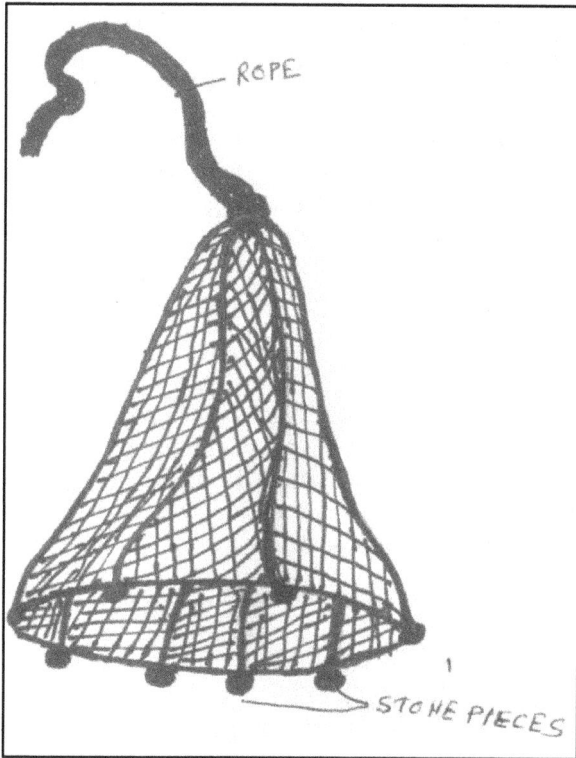

Figure 21.6: A Cast Net

varies from 1 to 5 cm (in descending order from the conical end) and the depth is between 2 to 5m. Sinkers in the form of iron rings are attached around the skirt. The hand rope is of cotton or nylon, 5-8mm in diameter and 7-15m in length. Generally the weight of the net is between 4.5 to 7 kg.

The net is cast into the water from the margin of the tank or pond or from a boat or stakes made of bamboo or wood in such a way that a group of fishes get covered over by the net and, thus, entrapped. The hand rope is carefully pulled to close the spread skirt. The overall result of operation is such that the fishes are caught in the pocket of the net. Its operation is always done in shallow water areas devoid of weeds and submerged obstructions of any sort. The fishing operation can be done all the year round. All types of fishes from surface feeders to bottom feeders get entrapped and brought to land.

Blanket Net

The net has a squarish piece of net webbing. It is suspended in water with ropes attached to its four corners, which are held on four boats. During day time some bait or during night attraction (light) is used to drive the fish over the suspended net. As the ropes are pulled the fishes get imprisoned in the bag of net and are collected.

Crafts used in Fisheries

Dugout Canoe

These are smaller types of boats and are used in the estuary, backwaters and rivers. These are cut out of a single log. These are known by different names like Danga, Dhoni, Vanchi, Hodi, Thoni and Ekhta. The size varies from about 3.5m to 14m in length. Generally the mango timber constructs it.

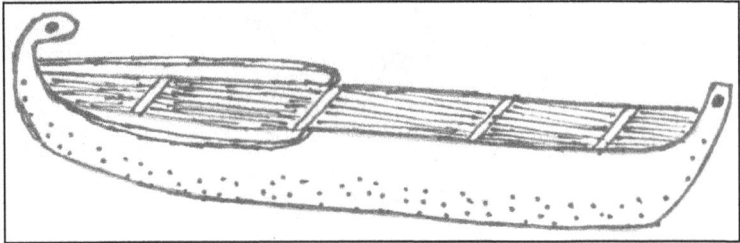

Figure 21.7: Odam–A Dugout Canoe of Malbar

Raft

It is a simple craft. These are also known as 'catamarans'. The name 'catamaran' is originated from the tamil word 'kattu maram' (kattu=binding, maram=words). It is a primitive craft, constructed by tying together several logs which are curved and shaped like a cone. One end of the craft is shaped into a cone, which rises above the water level and from the point from where rudder is controlled. The size of catamaran varies according to the number and size of logs used for its construction. The size of catamaran varies from 9-12 m in length, width 0.7to 1.4m and depth 0.3 to 0.7m.

Machwa

The size of this craft is about 9-14 meters in length. These are mostly used along Karwar on central west coast of India. It is mainly used for the transportation of cargo. In recent years these crafts are restricted themselves to some parts of west coast of India. It is a plank built craft. In recent years these have installed by inboard engines to enhance their speed and being used for fishing purposes also.

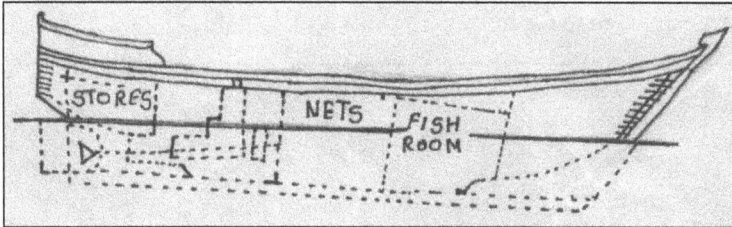

Figure 21.8: Fishing Machwa of Saurashtra

Masula

It is extensively used on the Coromandel coast and several Indian estuaries. It is about 9-12 meters or less in length. It is constructed with planks sown together with coir ropes and usually without frames or ribs. There are several variations of this type. In Orissa it is known as 'bar boats' and in Andhra Pradesh 'Padava' or 'Padagum'.

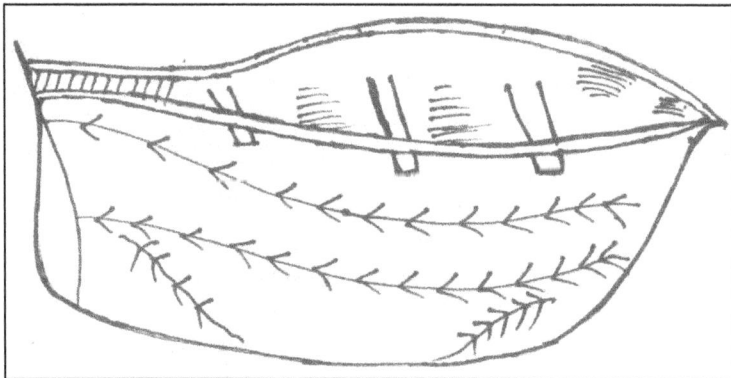

Figure 21.9: Padava–A Masula Boat of Andhra Coast

Satpati

It is also known as 'Galbati'. It has a medium pointed bow, broad beam, straight keel and high gunwale. Satpati has been found to be an ideal type for mechanization as a motor engine can be fitted without any change in the design of the locally assembled boat.

Tuticorin

It is a carved model. Its common size is 11x2x1m. These are also known as 'fishing luggers' and are used more as mother ships and cargo boats than for purpose of fishing.

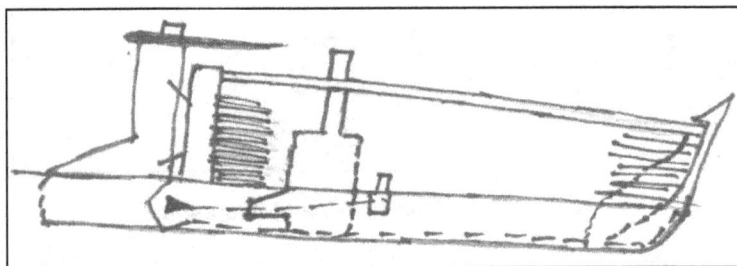

Figure 21.10: Tuticorin Type Fishing Boat of Tamil Nadu

Thermocoel Platform

The thermocoel platform of 6x3.5 feet size is used as fishing craft. It is covered by a plastic covering. This craft is not suitable to carry large nets which are heavy.

Coracle

It is a saucer shaped country craft, is also used in many Indian reservoirs. It is made of a split bamboo frame, covered with buffalo skin. Apart from being simple and inexpensive, coracle is durable. It is also a versatile craft used for laying and lifting of nets, besides navigation and transport of fish and other material.

References

Anon. 1980. Proc. 6th Workshop All India Cord. Proj. Ecol. and Fish. Freshwat. Reserv. Simla.

George, N. A., Khan, A. A. and Pandey, O. P. 1975. *Fishery Technology.* 12: 60.

Chapter 22

Fish Disease and their Control

Diseases in fishes are caused mainly by bacteria or fungal infection. Worms, copepods and various other parasites also infect various fish species. Some of the common diseases of the fish and their cure are discussed below:

Diseases Caused by Bacteria and Viruses

Bacteria and viruses affecting various organs of fish give rise to a number of diseases. Bacterial diseases are responsible for heavy mortality in wild and cultured fish. Attahment of the potential bacterial pathogen to the host's surfaces is extremely important for initiation of the infection. These pathogens come in contact with their potential host being carried by water or by direct contact or by their presence in the food. The gills, fins, gastro-intestinal tract, conjunctiva and skin surfaces are the common sites of attachment through which the organisms get entry into the body. Few important diseases are:

Dropsy

This disease is caused due to accumulation of liquid in some internal organs. The belly swells up considerably. The liquid accumulates in the belly. The intestine and liver are badly affected.

The disease-causing agent is a bacterium *Pseudomonas punctata*. It has been found that the dropsy is primarily due to virus infection, which is further complicated by a secondary infection of the bacterium. Infectious dropsy is one of the most feared diseases in carp culture.

Treatment

The fishes can be cured by treating with the antibiotic chloromycetin. The drug is dissolved in water (60 mg/gallon) and fish kept in it without food, till recovery, which may take 3-7 days. In acute cases liquid is first removed from the belly by means of a syringe before starting the treatment

Tail Rot or Fin Rot

This disease is caused by bacterial infection and results in the putrefaction of tail or other fins. The fins become torn and are gradually destroyed by the bacterial activity. A distinct white line is seen on the outer surface of fin in early stage of the disease. This line moves towards base of the fin and the fin becomes torn, and after sometimes the entire fin is completely destroyed. The infection may spread on body and invade the connective tissues

Treatment

a) Fish can cured by treating it with 1:2000 solution of $CuSO_4$ for 2 minutes. b) A bath in a dilute solution of acriflavine or phenoxethol has also reported effective against the fin rot. c) In case of serious infection the affected fin parts are removed surgically and the wounds are disinfected by touching with 1 per cent solution of silver nitrate. The fish is then placed in solution of 1:25000 of potassium dichromate for a week during which the wound heals up. The fin may be regenerated in course of time.

Furunculosis

This disease appears to infect fishes living in the dirty water containing a large amount of decaying matter. The bacterium responsible for the disease is the *Aeromonas salmonicida*. This disease is common in salmon and trouts. The characteristic feature is the formation of swellings or boils, which contain bloody, pus like substance, which may be discharged when the boil bursts. Severe infection results in the fish death.

Treatment

Treatment is done by removing infected fishes from the pond and by supplying food-containing antibiotics like nitrofurans and sulphonamids.

Eye Disease

An epidemic eye disease is caused by a bacterium *Aeromonas liquefaciens*. It affects eyes, optic nerves and brain of the fish.

During initial stages, cornea of eye becomes vacularized and later becomes opaque. Subsequently, the eye balls get putrefied leading to the death.

Treatment

During the initial stage of infection, chloromycetin @ 8-10 mg/l bath for one hour for 2-3 days, disinfection of pond with 1 mg/l potassium permagnate or 0.1 mg/l potassium permagnate followed by limning @ 300 mg/l.

Ulcer Disease

The causative bacterium of this disease is *Aeromonas hydrophila* and *Hemophilius piscium*.

The symptoms of the disease are ulcerations of the opercula and the head. Open sores or ulcers are present on the body of the fish, which increase in size gradually, exposing the muscles. Sometimes ulcerations are very deep and penetrate the cranial and opercular bones. In severe outbreak mortality occurs.

Treatment

1) Destroy the badly infected fish and disinfect the pond water with 0.5-mg/l solution of potassium permagnate. 2) Dip treatment for 1 minute in 1:2000 copper sulphate solution for 3-4 days in case of fish showing early stages of infection. 3) Oxytetracycline @25-75 mg/kg of fish per day with chloramphenicol for nitrite are effective. 4) Sulfonamide fed @ 3 gms per kg of feed for 12-20 days was reported to reduce mortalities in trout with ulcer disease.

Lymphocystis Disease

The causative agent is iridovirus. This disease is observed in most freshwater and saltwater species. The disease gains entry through epidermal abrasions. The virus infects dermal fibroblasts.

Symptoms

Clinically fish are seen with variably sized white to yellow cauliflower-like growths on the skin, fins, and occasionally on gills. Occasionally, this virus may go systemic with white nodules on the mesentery and peritoneum. Nodules may last for several months and cause infected fish to be susceptible to secondary bacterial infections. Reinfection can occur.

Tretment

The disease is self limiting and refractory to treatment (screen out the infected fishes as soon as possible to prevent the cross contamination).

Epithelioma Papillosum (Fish Pox)

The causative agent is *Herpesvirus cyprini*. It is nonfatal disease observed in carp and other cyprinids. There is epidermal hyperplasia with the epithelial cells occasionally demonstrating intranuclear inclusion bodies.

Symptoms

Elevation of the epidermis with the formation of white to yellow plaques over the body of the fish. Healed lesions usually turn black.

Treatment

Rearing fry and fingerlings in virus free water and hygiene controls.

Columnaris Disease or Cotton Wool Disease

Flexibacter columnaris and *Cytophaga columnaris* are responsible for this disease. The first external sign of columnaris disease is the appearance of grayish white spots on the fish body, often on the head, lips or fins. Some tissue –tufts or flakes of epithelium and bacteria are loosely attached to the mouths of fishes and appear like tiny wade of cotton, hence the disease is also known as 'cotton wool disease'.

In catfish, the initial lesions are smaller and circular with grayish blue centers and red margins surrounded by a ring of inflamed skin. Where the disease progresses, irregular slimy necrotic, grayish small flakes of tissue dangle in water.

In scaled fishes, the initial lesion appears on the outer margin fins as grayish white discolouration consisting of proliferating epithelium and bacterial cells. Gradually the lesion spread to the body and it appears to be carried with patches of mucus. A localized lesion often develops in the gills.

Treatment

1) Potassium permagnate bath at a concentration of 1:50,000 for 10-15 minutes; 2) Copper sulphate at a concentration of 1:30,000 for 20 minutes followed by placing the fish in running water; 3) Treatment of columnaris in aquarium fish can be done using brilliant green at 65 ppm for 45 seconds dipping, 20 ml of 3 per cent hydrogen peroxide in 1 liter of water for 10-15 minutes; 4) Chloramphenicol and chlortetracycline at 10-20ppm are found to be effective; 5) Topical application of $KmNo_4$ over the affected parts is a effective method of treating brood stock.

Fungal Diseases

Saproleginasis

This disease is also known as 'saprolaginasis' or 'cotton wool disease' or 'mold disease'. It is caused by Saprolegina parasitica. Infection caused by this parasite in fry and fingerlings of major carps is the main problem in fish farming. It develops not only in the diseased or dead fish but also develops in the weak and injured fish. The disease is characterized by a white to brown cotton like growth consisting of colonies of mycelium and filaments which appear as small to large patches on various parts of the body like fins, gills, mouth, eyes or muscle. The infection starts due to netting injury and overcrowding or lesions caused by other diseases. Saprolegina often infects the fertilized eggs. They attack the dead eggs first and then spread to the viable living eggs resulting in their spoilage.

The symptoms of the diseased fish are appearance of tufts of white hair in affected regions. In severe infection, fish are found rubbing against solid substrata. Fish becomes lethargic and dies after ulceration and haemorrhages.

Treatment

1) Dip treatment for 3 seconds in 1:10,000 solution of malachite green or for 5-10 minutes in 3 per cent common salt solution or 1:1000 solution of potassium permagnate is effective.

Gill Rot Disease

It is also called as 'Branchiomycosis'. It is caused by branchiomyces. The hyphae of these fungi grow into veins of the gills thus cutting off large areas of the gills from the circulating system. This disease is very common in over stocked ponds, having abundance of phytoplankton and organic matter.

Symptoms

The infected fish show damage of gill filaments, which become white and finally drop off. Fish gasps for air on the surface and dies of suffocation.

Treatment

Apply quick lime @ 200 kg/ha or copper sulphate @ 12 kg/ha in four monthly installments. Discontinue manuring and artificial feeding and add freshwater.

Diseases Caused by Protozoans

White Spot Disease

This disease is also known as ichthyophthiriasis or ich disease. Disease is very common among the freshwater fishes and is characterized by the appearance of white spots on the fins and the skin of the fish. Each spot is actually a small bladder containing the protozoan called 'Ichthyophthirius multifilis'.

The parasite is fairly large in size *i.e.*, about 1mm. It is spherical or oval in shape and is covered over by fine hair like cilia. The cytoplasm is granular and filled with fat like globules. The macronucleus is large and horse shoe shaped whereas the micronucleus is small and spherical. Micronucleus is situated near the macronucleus. Contractile vacuoles are present in the cytoplasm. The young parasite swims actively in water and on coming into contact with a fish, bores into the epidermis. Finally, it comes to rest between epidermis and dermis. Due to irritation a rapid proliferation of the epidermal cells takes place. The parasite then grows rapidly and appears as a small white spot. When fully grown, the parasite drops to the bottom and forms a cyst within which it multiplies rapidly forming a very large number of young ones. They are liberated from the cyst and attack a new host. The duration of life cycle of parasite is linked with the water temperature. High the temperature,

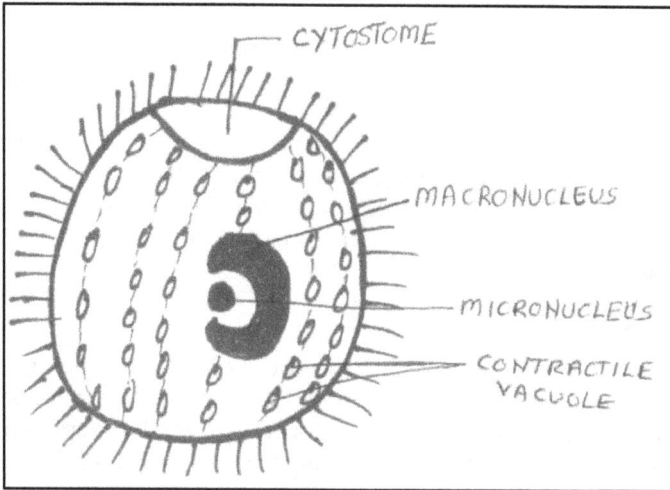

Figure 22.1: *Ichthyophthirius multifilis*

shorter the duration of life cycle and development. At lower temperatures, life cycle is prolonged and the ciliate remains on the fish for a much longer period causing the pathological damage.

Treatment

Control of this parasite is not easily possible by treatment with chemical, as it remains surrounded by the tissue of the host. Hence, the parasite can be killed when not embedded in the skin. Some of

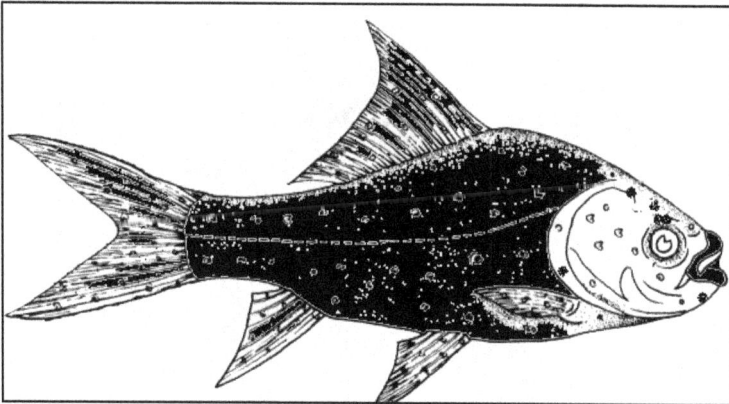

Figure 22.2: Fish Infected with White Spot Disease

the chemicals used to treat the disease are: 1) A 3 per cent salt solution or 1:4000 solution of formalin is quite effective; 2) A 1:100000 solution of quinine salt has been successfully used and it kills all the parities in one to two weeks; 3) Pure methylene blue in a solution of 1:500,000 may be added to the aquarium and the fish may remain in the solution for a sufficiently long time during which the parasites are killed, and 4) Persistent cases of white spots can be cured by treating with mepacrine hydrochloride. This drug is more poisonous than methylene blue.

Slimness of the Body

The causative parasite is *Chilodonella cyprini*. This parasite is common ectoparasite on the skin and gills of freshwater fishes in Asia, Europe and North America. It does not penetrate the epidermis. It is usually pathogenic only at temperatures below 10°C and is therefore, unlikely to be a problem in fishes in the aquariums. The ciliate is ovoid and dorso-ventrally flattened. It measures 50 to 70 mm long by 30 to 40 mm wide. Ciliation is incomplete. The ventral surface has 8 to 15 rows of cilia. The dorsal surface has a cross row of bristles. The cytostome is anterior and rounded. The oral basket is conspicuous and the macronucleus is oval. Transmission is by direct contact with the free swimming organism in the water.

Treatment

This disease can be controlled by treating the infected fish with

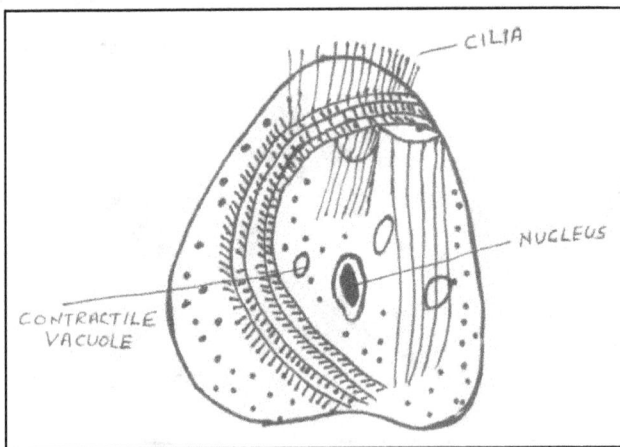

Figure 22.3: *Chilodonella cyprini*

1:4000 solution of formalin for one hour.1:500 solution of acetic acid is also effective in killing the parasites.

Costiasis

Costiasis is caused by a protozoan ectoparasite called costia. It may be fatal. The symptom of the disease is the appearance of light bluish or grayish film of mucus over the body of the fish. The fish loses appetite and becomes weak, then finally dies after some time. This parasite has groove leading into the gullet and bears two pairs of flagella. The parasites live on the skin or gills of the fish where they destroy the epidermal cells and feed on

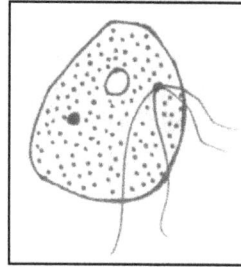

Figure 22.4: Costia

them. This disease is associated with the pond fishes living densely in water having low pH and poor conditions of food. The parasite attacks the fishes of all the ages and live in large numbers on their skin, fins and gills.

Treatment

1) to keep fish for 10 minutes in either 3 per cent of common salt solution or in 1 to 2500 formalin solution. Overcrowding should be avoided and pH of water should be checked regularly.

Myxobolus

It produces the disease called 'whirling disease' or 'myxosporidian disease'. It produces cysts in different regions of the body and internal tissues and organs of the fish. This disease is common in mrigal and American salmonids.

The symptoms of the disease are small white cysts over the body especially on gills and fin bases, tail chasing behavior while feeding, weakness, emaciation, raising of scales along their posterior margins, falling of scales, loss of chromatophores, perforation of scales, small white spots, restlessness etc.

Treatment

The diseased fish should be disinfected with a weak solution of sodium chloride. Thinning of fish population and using yeast pellets @ 1 gm/kg of feed are also effective.

Trichodiniasis

This disease is caused by *Trichodina indica*. The carp fry and fingerlings sometime carry heavy infection of the gills and skin with *Trichodina indica*. In heavy infection there is excessive mucus secretion on the gills which hampers respiration and the fish start surfacing. Some species of Trichodina occur inside the urinary bladder. Under microscope they appear as circular transparent organisms with an internal disc like structure. Fish affected with trichodiniasis turn sluggish, lose weight and become moribund. There may be excessive mucus secretions and epidermal necrosis.

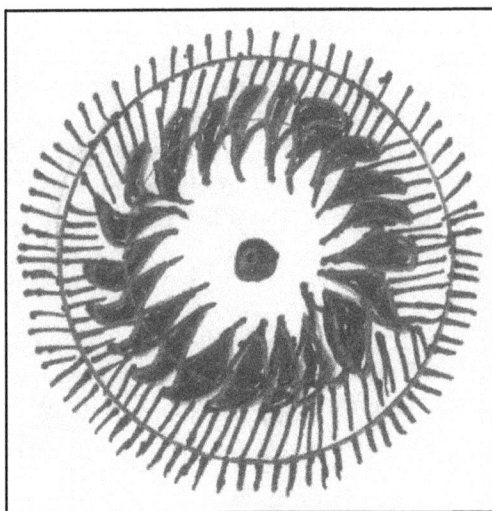

Figure 22.5: Trichodina

Treatment

A short bath in 2 to 3 per cent salt solution for 1-2 minutes can control the infection.

Diseases Caused by Worms

Two types of flukes chiefly parasitise the fishes. The monogenetic and the digenetic. Monogenetic flukes complete their life cycle within a single host and then transferred directly from one to another fish. They are represented by about 6000 mostly the host specific species, which attach on external body surface of the fish and cause heavy mortality.

Digenetic flukes complete their life cycle in two hosts. If the hosts other than the fish (secondary hosts) are destroyed their life cycle never get complete. Their control therefore becomes easier. They are primarily external and become internal only secondarily.

Gyrodactylus

It is very common on gills and skin of the freshwater fish. It infects the trouts and carps. The affected surface becomes covered over by bluish slime due to increased secretion of mucus. When the infection is heavy, colour of the fish fades and becomes pale. The skin becomes slimier; the fin becomes frayed and gradually becomes torn. Gyrodactylus can be easily seen by scraping mucus and examining it under the micrscope. Fish infected with the parasite are often seen rubbing their body against the sides or bottom of the aquarium or pond to get rid of the parasite.

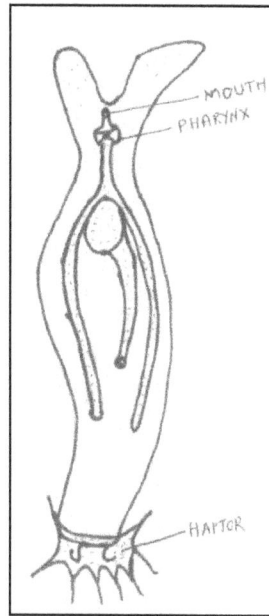

Figure 22.6:
Gyrodactylus

Gyrodactylus has two conical projections at the anterior end, which bear openings of glands producing sticky liquid to help them to adhere to the skin or gill of the fish. At the hind end is a strong disc like organ of attachment bearing hooks and is called as 'haptor'. Eyes are absent. It does not lay eggs but gives birth to living youngones, which can be seen within the body of the mother in various stages of development. Another embryo may be present within the body of the unborn youngone. The reproduction probably takes place by sexual process, as well as by paedogenesis.

Treatment: A good treatment is to place the fish for 1 or 2 minutes in 1:500 solution acetic acid or by placing fish for 5-10 minutes in 1:2000 solution of ammonia. The best method is to treat the fish for a longer time (2-3 days) in 1:4000 solution of formalin).

Dactylogyrus

It is found only on gills of fishes. The affected surface becomes covered over by bluish slime due to increased secretion of mucus.

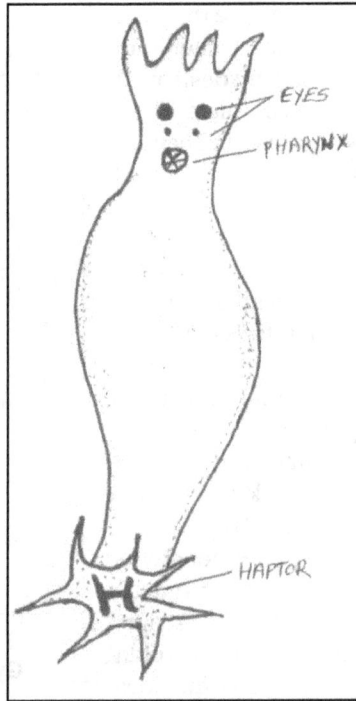

Figure 22.7: Dactylogyrus

When the infection is heavy, the colour of fish fades and becomes pale. It has two conical projections at the anterior end which openings of glands producing sticky liquid to help them to adhere to the gill of the fish. At the hind end there is a disc like strong organ of attachment called 'haptor'. Haptor bears hooks. A pair of eyes is present at the anterior end. It lays eggs.

Treatment

A good treatment is to place the fish for 1 or 2 minutes in 1:500 solution acetic acid or by placing fish for 5-10 minutes in 1:2000 solution of ammonia. The best method is to treat the fish for a longer time (2-3 days) in 1:4000 solution of formalin).

Diplostomiasis

It is caused by digenetic trematode called 'diplostomium'. This disease is more prevalent in culture ponds. Numerous small nodules or cysts of about 1.3mm diameter appear all over the body of the fish.

Treatment

Isolating the infected specimens and giving them dip treatment in 3:1,00,000 picric acid for a period of one hour.

Pisciola

It feed on the blood of the host fish causing irritation and abnormal movements. The wounds made by leeches are affected by fungus, causing more harm to the host.

Treatment

Dip treatment in 1:1000 solution of glacial acetic acid disinfections with 1:10,000 potassium permagnate solution. A single treatment with 5 mg/l Gammexene was reported to cure fishes in sewage fed ponds. Dip treatment in 2.5 per cent common salt solution for about 30 minutes is also advisable.

Diseases Caused by Crustaceans

Argulus

It is commonly known as fish louse. This is a well-known fish parasite and is widely distributed on many fishes. It remains attached

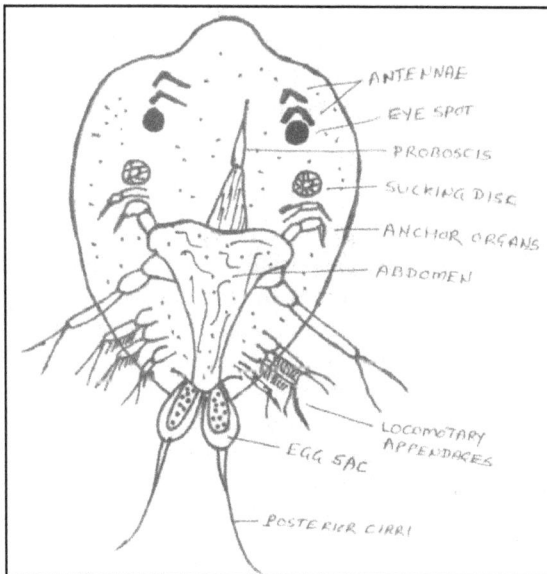

Figure 22.8: Argulus (Female)

to skin, fins and gills of host by means of two large suckers. These parasites cause extensive pathological lesions in the skin showing circular haemorrhagic patches which become ulcerated. These lesions promote secondary infections like fungi and bacteria. Rohu has been observed to be the most susceptible species among the Indian major carps. Skin from the parasitized areas show epithelial degeneration, oedema and hyperplasia of dense connective tissue. Kidneys exhibit marked glomerular changes, tubular degeneration and necrosis. In the liver, increased sinusoidal spaces and focal necrotic areas are also noticed. Gill lamellae show hypertrophy and hyperplasia in many areas resulting in the fusion of secondary lamellae. The female louse dies after spawning and the eggs are deposited on stones or on the glass of aquarium. The eggs hatch after about one month and the young one undergoes moulting 4 or 5 times, after which the young parasite is capable of attacking a fish.

Treatment

(*a*) As the fish louse is quite large, it can be easily removed by means of a forcep. Parasites that are strongly attached to the host, can be paralysed by touching with a strong salt solution and then removed (*b*) In large ponds $kmno_4$ may be added (0.2 to 0.3 grain/ gallon of water). Too strong solution is dangerous for the fish (*c*) Application of malathion @0.25 ppm for three consecutive doses (*d*) Application of cypermethrin in the pond with judicious dosing is more effective than other chemicals.

Anchor Worm (Lernaea)

It is a copepod crustacean, which resembles a worm due to the loss of appendages. This parasite is reported to infect fingerlings and adults of major carps especially catla. The female penetrates deep into the flesh and becomes attached by hook like appendages at the anterior end. The male does not attack the fish and is different in shape. This parasite penetrates the muscles and reaches the blood vessels as it feeds on the host's blood. The fish thus become anaemic and weak. The parasite causes wounds and holes on the body, which become infected with bacteria and fungus and the fish dies from these secondary infections.

Ergasilus

This parasite infect gill-lamellae. They attach themselves to the gills with the help of their hook like second antanna. Due to this

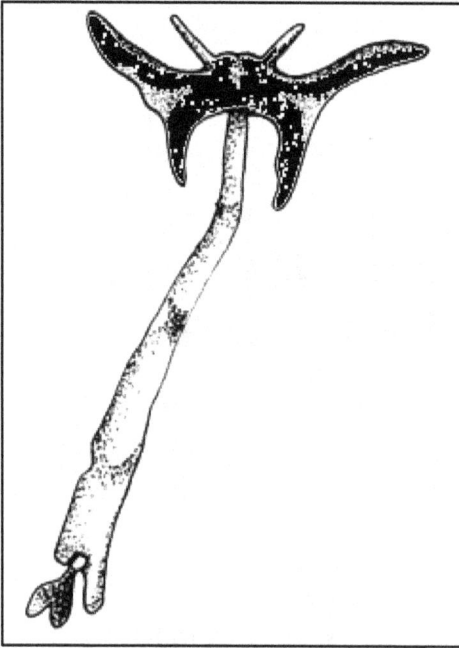

Figure 22.9: Lernaea   **Figure 22.10: Ergasilus**

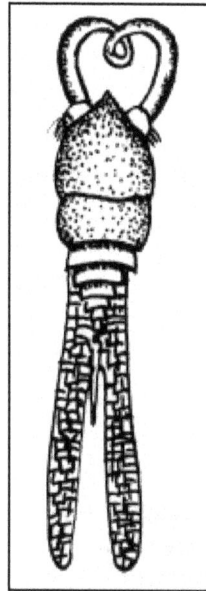

there are wounds and haemorrhage in the gill tissue. There are no reports in India where mortality of fish has taken place due to this parasite. The infected fish suffers from anemia, respiratory difficulties and poor growth. Further complications appear due to secondary bacterial and other infections.

Chapter 23
Sewage-Fed Fish Culture

Sewage is a rich nutrient resource, cheaply available around big towns and cities. It contains near about 99 per cent of water and 1 per cent of other materials including bacteria, pathogens and various organic and inorganic substances present in the true molecular, colloidal or semi colloidal state. It is universally considered as a valuable organic fertilizer as it contains abundant quantities of nutrient elements like phosphorus and nitrogen. The use of sewage effluent for fish culture has been recognized in many countries. It is widely used, to increase fish yield in the countries like Germany, USSR, Poland, Hungary, Israel, China, Japan and Indonesia. In India the use of sewage in fish culture is an age-old practice. Initially the chief fisheries were located in West Bengal and Madras, but now with the use of sewage treatment, its application in profitable fish raising is realized in the states like Bihar, Uttar Pradesh, Maharashtra, Madhya Pradesh, Tamil Nadu, Karnataka and Kerala. The fish farmer of Kolkata developed a unique technique of utilization of domestic sewage for fish culture long back in 1930s. The early inspiration of utilizing the sewage for fish culture emerged from the waste. Stabilization pond used as water source of vegetable fields. This technique is considered to be the largest operational system in the world to convert the waste in consumable product. The growing fish demand of the metro city Kolkata is widely met by this technique.

The area under sewage-fed fish culture reached up to 12,000 ha. But recently due to rapid and indiscriminate urbanization it has come down to 4,000 ha. (approx) resulting in crisis of livelihood of rural people.

Characteristics of Sewage

The domestic sewage contains 99 per cent of water. Other substances constituting the remaining 1 per cent include various organic and inorganic substances, chiefly the carbon, nitrogen, phosphrous and the heavy metals. The gaseous component of sewage includes ammonia (NH_3), carbon dioxide (CO_2), and hydrogen sulphide (H_2S) gases. Heavy metals comprise the traces of manganese, chromium, copper, zinc and nickel etc. Besides, the detergents, bacteria and disease causing microbes also form a variable proportion of the sewage.

The raw sewage is not directly supplied to the fishpond, because it has high biochemical oxygen demand (approximately 200 ppm), low dissolved oxygen contents, high carbon dioxide contents, high values of ammonia, hydrogen sulphide and the bacterial load. All the aforesaid characteristics necessitate a pretreatment of sewage, to make it useful in fish culture. Properly treated sewage when diluted in right proportions with the freshwater and supplied to pond, beneficially influences the fish culture in the following ways:

1. During the treatment, the organic matter undergoes bacterial decomposition and a large amount of CO_2 is produced. This gas is used up by the algae, which as a result flourish and produce algal blooms. These blooms and algae are consumed in large amounts by fry, fingerlings and adult stages of several fishes under cultivation.

2. As the algae, as primary producers flourish, the zooplanktons develop forming the first trophic level of consumers. These along with the algae are consumed as food by the fishes.

3. The larger particles of sewage are consumed directly by the fishes and those present in soluble form are utilized by phytoplankton and zooplankton for their growth.

4. The inorganic nutrients released as a result of microbial decomposition of dissolved organic matter add up further to the fertility of the pond.

Sewage Treatment

Owing to lack of oxygen, the sewage is not suitable directly for fish culture. Hence before letting it out to fishponds it is to be treated by any of the following methods:

Mechanical Treatment

Mechanical process consists of screening and filtration, so as to remove coarse suspended matter. Floating solids including fats and oils, as well as fine suspended material are removed by skimming and sedimentation. Sedimentation is done by letting the sewage into a tank at a high velocity. When the sewage enters a large tank from a sewage channel, there is sudden drop of velocity, resulting in sedimentation.

Chemical Treatment

Different chemical methods include the deodorization of sewage by chlorine and ferric chloride; the sterilization or disinfection by using copper sulphate and chlorine and coagulation or precipitation by adding certain coagulants like ferric chloride, lime, alumn and organic polymers to the sewage.

Biological Treatment

Biological treatment involves bacterial decomposition of the organic contents of sewage into H_2O, CO_2, nitrogen oxide and sulphate etc. It is accomplished by the bacteria which decompose the substances either anaerobically or aerobically.

General Method

From the viewpoint of fish culture the sewage is more commonly subjected to three kinds of treatments *viz.* sedimentation, dilution and storage.

Sedimentation

The function of sedimentation is to remove suspended solids from sewage to the maximum possible extent. It is done by letting sewage into a pond/tank at a high velocity of flow. The process is generally completed in two sedimentation tanks, of which the first is built at a level higher than the second.

In first tank the sewage is retained for about 10-12 days for initial sedimentation. During this period of stagnation most of its particulate matter settle down and organic matters are decomposed aerobically to produce the inorganic nutrients.

The second tank is known as the waste stabilization tank. It is built slightly below the level of the first. It receives sewage from the first tank that flows and then fall at high velocity into it. Sudden drop in the velocity of sewage in the second tank causes further sedimentation of coarse particles and also the homogenization of the sewage variations. The sewage in this tank is allowed to stay for 20 days, during which the water losses its foul odour and becomes rich in algal blooms.

The stabilization tanks are categorized into three types *viz.*, the aerobic, facultative and the anaerobic tanks. The aerobic stabilization tank is the shallowest being only about 0.5m deep. In this tank the decomposition takes place aerobically to maximize conversion of organic matters into stable ones, consequently the algae grow to the maximum extent. Facultative stabilization tank is about 0.9 to 1.5 m deep. It is aerobic in day time and becomes anaerobic during the night. Anaerobic stabilization tank is about 2.5 to 3.7 m deep. In such a tank, the organic substances are degraded anaerobically by two types of bacteria. The first type of bacteria hydrolyzes and ferment the complex organic substances into primary volatile acids and alcohol by the process called the liquification and microorganisms accomplishing the process called 'acid fermenters'. The second group of bacteria converts the produced acids (*e. g.*, acetic acid, propinoc acid etc) and alcohols into CO_2 and methane gases. This process is known as gasification and microorganism involved are called as 'methane formers'.

The size of oxidation tank varies. Oxidation tank is selected according to input and outflow of sewage from it, keeping in view that the sewage can be discharged from it for 10 days continuously. For a daily inflow of one lakh liter of sewage, a pond of 50x20x1.5 m is suggested. Above the desired depth of tank a small portion is generally left to adjust variations of the daily input in the oxidation tank.

Dilution

The odourless stabilized sewage flows out of second tank, through an overflow channel. Now it is diluted with freshwater in a ratio of 1:5 (one part stabilized sewage and five parts of water) to restore the amount of dissolved oxygen, and to reduce concentration of substances like CO_2, NH_3, H_2S etc, below the lethal limits of fish selected for cultivation. By adding freshwater O_2 level is regained and improved. Further by photosynthetic activity by green algae and aquatic vegetation.

Diluted sewage is supplied to the ponds, where the carp fry consume the algal blooms and attain stocking sizes within a few days. Sewage water added to stocking pond is beneficial for the fish growth because it is rich in fertilizers and food, particularly for the herbivorous fishes.

Fishes Cultivated

As the stabilization ponds are poor in dissolved oxygen contents, the air breathing fishes are only selected for culture in such ponds. The fishes such as *Catla catla, Labeo rohita, Cirrhinus mrigala, Tilapia mossambica, Clarias batrachus, Heteropnestus fossilis, Channa* spp. and *Ctenopharyngodon idella, Cyprinus carpio* and *Hypophthalmichthys molitrix* can survive and grow well in these ponds.

Ghosh *et al.* (1973) investigated the survival and growth of the silver carp by rearing this fish in sewage fed pond at Khardah in West Bengal. He reared a population of silver carp in nursery ponds for about one month. The fry of average 35mm length attained a length of 110mm and weight of 8gm in this short period. The grown-up fries were then transformed to the sewage fed stocking pond wherein other major carps, such as rohu, catla, and mrigala were also stocked in the ratio of 2:5:3. The silver carp introduced were only 65, which contributed 0.2 per cent of the total stocked population.

Tilapia has been cultured in Great Britian, Germany, Russia and the United States. The nickname 'the aquatic chicken' has been used on tilapia, due to its ability to grow quickly with poor quality inputs. Tilapia is hardy and is most suited for sewage irrigated pond culture. Their fast growth rate, ease of breeding, lesser demand of dissolved oxygen and above all the survival at highest attainable ammonical nitrogen level of 5.43 ppm, make them a good choice for fish farmers. These fishes mature at a small size and begin to reproduce earlier before they grow in size. It is therefore suggested to either raise the opposite sexes in different ponds or to introduce catfishes to control the population of the fry. Ghosh *et al.* (1976) reported a total production of 220 kg/hectare area, in a polyculture of tilapia and clarias. In another experiment Ghosh *et al.* (1979) utilized a sewage-irrigated pond (0.076 ha) for massive tilapia production. Tilapia were not found to be affected even at the highest attained NH3- N level of 5.43 ppm. In this practice they got the production of 9350 kg/ha/yr. It is thus evident that the culture of tilapia in both the mono or polyculture systems gives best cost effect of their raising.

Clarias batrachus (magur) grows very fast in sewage waters or oxidation pond. Hence it may be cultured in sewage fed pond with advantage. Magur showed a remarkable growth of 195.9 gm from the initial stocking size of 15gm within a period of 100 days while cultured with tilapia in a sewage fed pond. Thus the sewage fed ponds have very high productivity with no cost of pond fertilization and supplementary feeding (Pal, 1978).

Carps are very sensitive to dissolved oxygen content of water. They are thus raised in ponds receiving diluted sewage water. A total production of 6 ton carps per hectare area has been obtained so far in treated sewage fed waters. The fingerlings of *Cyprinus carpio* stocked in primary oxidation pond, at a rate of 80 kg per hectare were recorded to give an annual yield of 11938 kg per hectare (Akolkar and Belsare, 1984).

As the sewage contains high content of nutrient, the farmers keep very high stocking density. The stocking densities for catla, rohu, and mrigala, stocked together in treated sewage fed water is suggested to be 20,000 per hectare in a ratio of 1:2:1 stocked with other carps is suggested to be 10,000 per hectare in the following ratio:

1. Silver carp, catla, rohu, mrigala, common carp, and grass carp in a ratio of 25:15:10:25:20:5

2. Catla, rohu, mrigala, common carp, and fringe lipped carp in a ratio of 40:10:20:20:10.

Air breathing murrels like *Channa sp* and catfishes like *Heteropneustes fossilis* may also be grown along with the carps in sewage fish culture. However, they should be introduced only after the fingerlings of carps have grown sufficiently big in size, so that they may not fall victim of these fishes.

From the above account, it can be well concluded that the sewage fish culture is profitable chiefly because it does not require fertilizers and supplemental feed, both reducing the cost of production and also because that in sewage system fish growth is relatively higher. Recently a high yield of fresh water prawn (*Macrobrachium rosenbergii*) is recorded from the paddy fields supplied with the sewage water.

The freshwater carps such as catla, rohu, common carp and silver carp are suitable for human waste fed fish culture. Considering the rich nutrient status of human wastes, a fish yield of 6t/ha can be obtained by stocking fingerlings at 20,000/ha. The revenue through human waste-fed fish culture can be further augmented; if the excess treated waste is used for irrigation of fodder crops, vegetable crops, coconut, palm etc. To ensure greater safety of fish flesh free of pathogens, the waste-grown and harvested fish may be maintained for a week in clean water, which may eliminate the disease causing organisms that survived the easier treatments. Proper cooking of the fish further renders the fish completely safe for human consumption.

Disease Management

The fishes are most vulnerable to bacterial diseases, but surprisingly the occurrence of bacterial or any other disease is not common in sewage-fed fish farms. Even when Epizootic Ulcerative Syndrome was prevailing in recent years in other areas, the sewage-fed ponds were uninfected. However, parasitic infections by Lernea and Argulus are common.

References

Balkrishna, Gopal. 2007. Sewage-fed aquaculture-A biological method of waste treatment. Seminar write up submitted for partial fulfillment of Post Graduate Diploma in Inland Fisheries, Central Institute of Fisheries Education, Salt Lake, Kolkata Centre,

Ghosh, Apurba, Banerjee, M. K. and Hanumantha Rao, L. 1973. Some observations on the cultural prospects of silver carp, *Hypopthalmichthys molitrix* (Val.) and sewage fed ponds. J. Inland Fish. Soc. Ind., V: 131-133.

Ghosh, Apurba, Banerjee, M. K. and Saha, S. K. 1979. Cultural prospects of *Tilapia mossambica* Peters in a pond fertilized with domestic sewage (Abstracts), February 12-14, 1979, Central Inland Fisheries Research Institute, Barrackpore.

Chapter 24

Larvivorous Fishes

Atkin (1901) asserted that few fishes consume mosquito larvae as their food. If these fishes are introduced into the water, they will feed upon the population of mosquito larvae, which are the vector of several diseases like yellow fever, margina, malaria etc. These fishes breed in standing water and shallow weed infested water bodies and all other kinds of inland water bodies. The different types of insecticides and DDT are also used to control the mosquito population, but these chemicals have their ill effects on human health also. Fish is a natural enemy of mosquito eggs and larvae, and its use as a means of biological control has been recognized since olden times.

Characteristics of Larvicidal Fishes

A larvicidal fish should possess the following features:

1. It should be small so as to be capable of moving freely among the weeds.
2. It should be a hardy fish
3. It should have little food value though of immense value as larvicidal fish.
4. It should be easily available
5. It should breed freely in confined waters
6. It should be a surface feeder and carnivorous in habit

With such characteristics the fishes may be used as substitute of insecticides in carrying out the biological control of vector insects. A list of different indigenous and exotic larvicidal fishes of India is given below. From these the fishes may be suitably selected for control of the mosquitoes.

Indigenous Larvicidal Fishes

Notopterus notopterus, Oxygaster, Danio, Barilius, Laubuca, Rasbora, Esomus danricus, Puntius, Oryzias melanostigma, Ambassis, Mugil, Etroplus, Anabus, Therapon, Badis, Glossogobius, Aplocheilus lineatum

Exotic Larvicidal Fishes

Gambussia affinis, Lebistes marginated, and *Carassius auratus*

Following is a brief account of some larvicidal fishes in India:

Indigenous Larvicidal Fishes

Notopterus notopterus

It is commonly known as 'chital'. Body is strongly compressed with a short pre-caudal region. Head compressed with moderate mouth. Dorsal fin inserted nearer snout tip than to base of caudal fin. Pectoral fins moderate, extend beyond anal fin origin. Scales minute. Colour is silvery dark greenish or glossy yellow on the back. The fish is found in fresh and brackish waters. The youngones of the fish feeds upon the mosquito larvae, but the adult feeds upon the small fishes. As the fish consumes other small fishes, it is unsuitable

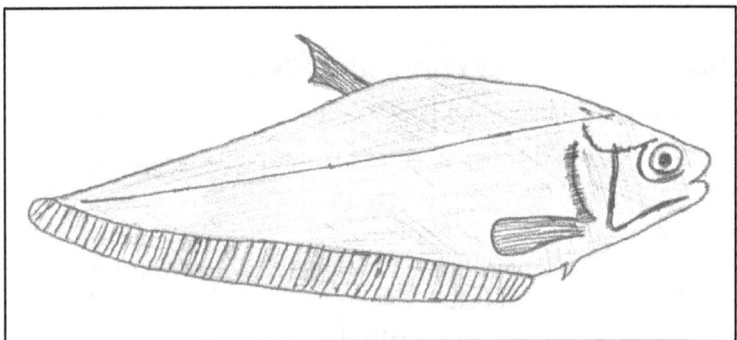

Figure 24.1: *Notopterus notopterus*

as larvicidal fish. It inhabits fresh and brackish waters. It is widely distributed in India, Nepal, Thailand, Indonesia, Pakistan and Bangladesh.

Mugil cephalus

It is commonly found in brackishwater and backwater. It also grows well in freshwater ponds. Body is short and flattened with a broad terminal mouth. Upper lip is thin and smooth. Among two dorsal fins, the first is short with 4 slender spines, and it arises midway between the end of the base of the caudal fin. The second dorsal fin is usually with 8 soft rays. Pectoral fins are situated above the middle of the depth of body. Anal fin originates opposite to the second dorsal. Caudal fin is with pointed lobes. A dorsal black spot is present on the pectoral base. Colour of the body is grayish along the black and belly is often silvery with grey stripes lengthwise. The youngones are surface feeder and feed chiefly on the mosquito larvae. Adults are valued as food fishes. Fish attains sexual maturity at 25-50 cm length weighing 1.2 to 2.0 kg.

Chela labuca

It is commonly known as Indian glass-barb or Indian hatchet fish. Body deep and compressed. Mouth slightly oblique. Pectoral fins large and wing like. Lateral line complete. Body colour shiny silver to greenish-grey with a violet luster on caudal peduncle and steel-blue vertical markings on sides of the body. It is widely distributed in Pakistan, Bangladesh, Sri Lanka, Burma and India. This species is useful substitute and is of fair quality for larvicidal work.

Esomus danricus

It is a fast swimmer. Body elongated, slim and compressed. Mouth is small. Mouth opening is obliquely directed upwards and is small. Lips are thin and lower jaw is prominent. It has greenish body with a steel blue band in the flanks and a black streak along the dorsal surface. The maxillary barbel extends up to the middle of the pectoral fin. Being very delicate, they lose scales and may die immediately after catching, if special care is not taken. Caudal fin forked and eyes are placed inferiorly, visible from below ventral surface. Pectoral fins pointed and lateral line incomplete. They always

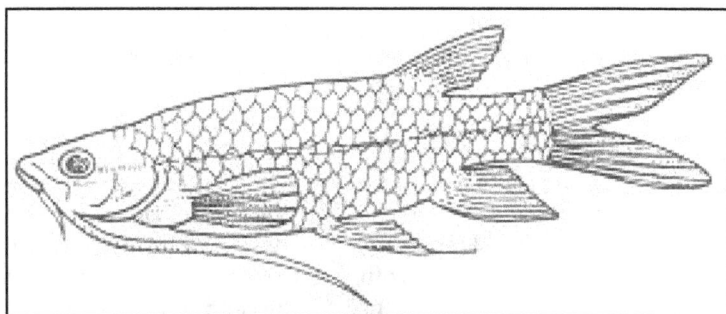

Figure 24.2: *Esomus danricus*

remain in the surface layers of the water column. Its eggs are partially adhesive. It attains about 5 inches in length. It is widely distributed in Pakistan, Nepal, India and Sri Lanka. It remains in the upper layers of the water column. It is considered effective for mosquito control.

Etroplus suratensis

Body is oblong and compressed. The lower jaw is slightly longer than the upper jaw, the cleft of mouth is small. Teeth are compressed and found in a single row. Dorsal fin is single with its spinous portion much larger than soft portion and there are 18-19 spines and 14-15 soft rays. Caudal fin is slightly emarginate. Body colour is light green with 8 vertical bands of blackish and slate colour. Advanced fry feeds on aquatic insect larvae. Adults are herbivorus. The eggs are brownish in colour. It occurs in estuaries, backwaters and creeks of Indian peninsula and also in freshwater ponds.

Chanda ranga

It is commonly known as Indian Glass fish. The body is deep and rhomboidal in shape. The eyes are large. The dorsal fins are two in number. The body is transparent and glass like. The internal organs are clearly visible from outside attesting the transparency of the fish. These organs are packed in a small silvery sac in the abdomen. Males are having bigger fins than the females. The male has a light blue trailing edge to its dorsal fin. It grows up to 3 inches. It mainly feeds on crustaceans, but can be used for mosquito control.

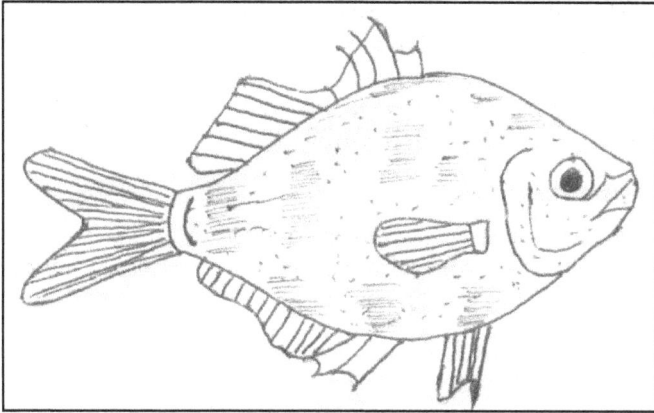

Figure 24.3: *Chanda ranga*

Aplocheilus panchax

It is commonly known as 'Blue panchax'. It is widely distributed in Eastern India, Thailand and Malaysia. The body colour is olive in different shades. The rounded tail is bordered with reddish, yellowish and blackish colour. The females are rather colourless. It is a hardy fish and grows to a maximum length of 9.0 cm.

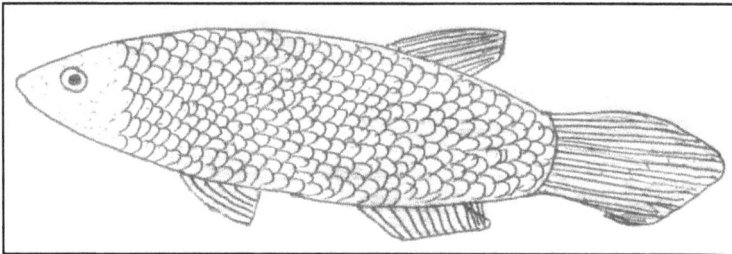

Figure 24.4: *Aplocheilus panchax*

Colisa fasciatus

It is commonly known as 'Giant Gourami'. Body is elongated, fairly deep and laterally compressed. It has a red brown coloured body with transverse blue green iridescent bands on the body. Pelvic fins are long thread like which can moved in all directions. The blue anal fin is bordered with red and blue dorsal fin having red spots scattered on a blue back ground. The dorsal, anal and caudal fins are with blue flanks. Ventral fins are orange red. Eggs are lighter and floating type. The fish can grow up to 12 cm.

Rasbora daniconius

It is hardy and efficient larvicidal fish. A bluish black lateral stripe along the centre of the slim body that extends from the tip of the snout to the tail. It is gregarious and active peaceful fish.

Oxygaster bacaila

It is commonly known as 'chelva'. It is of great utility in mosquito control. Fish feeds on mosquito larvae throughout life. Mouth is upturned. It is surface feeder. Larger species are sued as food.

Puntius ticto

It is commonly known as ticto barb or firefin barb. It is small sized hardy fish. Body elongated and colour is black grey to grassy green. Long transverse black blotch above the pectoral fin and another similar on the caudal peduncle over the end of anal fin. Mouth terminal and small. Barbels absent. Dorsal fin inserted slightly posterior to pelvic fin origin. The dorsal fin of the female is pale, except for a faint rose at the breeding time. Lateral line complete. It is distributed in India. Pakistan, Nepal, Thailand and Sri Lanka. It attains a length of 10 cm.

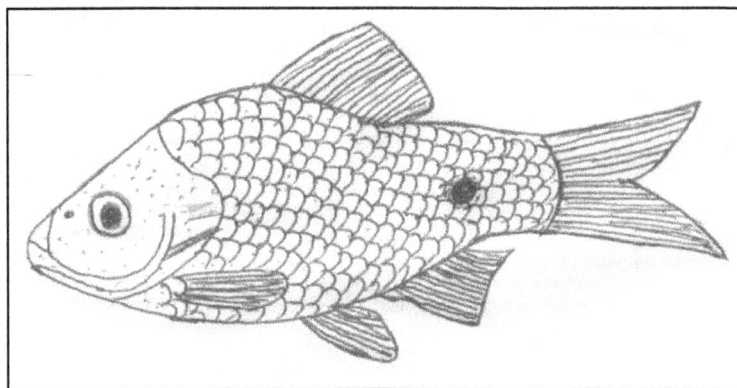

Figure 24.5: *Puntius ticto*

Oryzias melanostigma

This is a euryhaline fish which spawns in shallow water. It cannot stand stress of transportation. Hence, it is suggested to use it as an efficient destroyer of mosquito larvae, in the region where it is found. It is 1.5 inch in length.

Danio rerio

It is a small sized fish which grow up to 5.2 cm in length. It is with upturned mouth and considered as a good larvicidal fish. Body is elongated and strongly compressed. Mouth terminal and oblique. Anal fin long and caudal fin forked. Pectoral fins much longer. Lateral line complete.

Exotic Larvicidal Fishes

Carassius auratus

There are more than 20 varieties of fancy breeds including fan tail and veil tail gold fishes. The common goldfish has an elongated, stock body. The basic body colour is greenish-gray to orange with a variable iridescence in reflected light. Gold fishes are mostly golden yellow, however the colour varies from grayish silver to all black. In natural habitat, it reaches up to 45 cms. These fishes are found in still waters with plenty of vegetation. It is omnivorous, mainly feeds on plant matter. Gold fishes sexually mature within 5-6 months. The males develop breeding tubercles during spawning period. It is prolific breeder. Eggs are adhesive which attach on the weeds and the larvae hatch out. This ornamental fish can be used for mosquito control. But due to its high cost, its use on a large scale is uneconomical.

Poecilia reticulata

It is commonly known as 'Guppy'. The fish has been named after an aquarist, Rober John Lechmere Guppy who had sent a few living pairs of this species to the British Museum, London in 1860. As guppies are collected in large numbers from freshwater bodies, they also bear the name 'million fish'. Though the guppies had their origin in West Indies, they are grown in many countries as ornamental and larvicidal fish. Guppies are small fishes with a variety of caudal fin evolved through crossing and selection. It was introduced in India from South America in 1908. In India, it is now

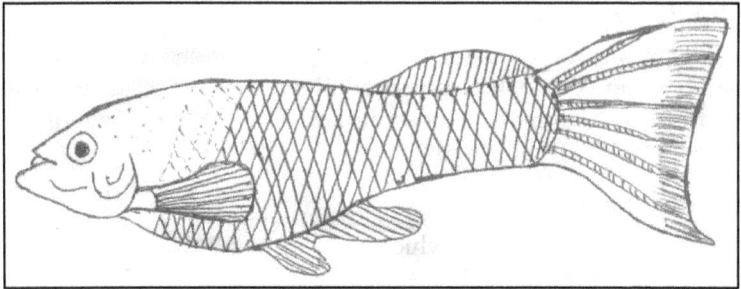

Figure 24.6: *Poecilia reticulata*

well established in certain parts of Tamil Nadu, Andhra Pradesh, West Bengal, Orissa and Maharashtra.

Body cylindrical with moderate mouth. Small teeth present on jaws. The dorsal fin is irregular and the eyes are large. Mouth is upturned. The dorsal fin is irregular. The common colours are red, green, blue, grey and yellow with spots. Young males are easily distinguished from the females by their beautiful orange, red and black dots all over the body and fins. The females are olivaceous. Males are always smaller than females. Gravid females can easily be recognized by the young they contain and the adult males are marked by the elongated anal fin which is used as intermittant organ by the male to guide spermatozoa into the females. Males grow up to 3 cm and a female up to 5 cm. It is omnivorous and voracious feeder. Female releases 15-25 youngones at a time. The optimum water temperature range for breeding is 20-24°C.

Gambuisa affinis

It is commonly called as 'top minnow'. This fish has been extensively applied for mosquito control in India as well as abroad.

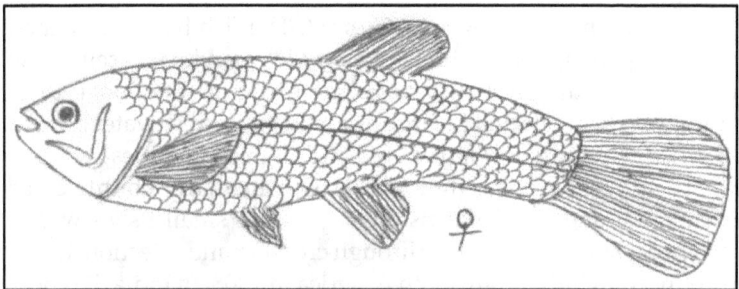

Figure 24.7: *Gambuisa affinis*

Brought from Italy it was introduced in India in 1928. It is highly voracious and destructive fish. The fish can grow up to 9 cm. It breeds at 1.5 inches. The females are larger in size than the males. Males are provided with gonopodium. This fish may consume about 170 larvae in 12 hours. It is extremely hardy fish.

Chapter 25
Aquariculture

Ancient Egyptians were the first humans known to keep fish not only for food purposes but as a source of food and entertainment. In 1853, The London Zoological Society established a public aquarium to display fish species. By 1864, public aquariums had been opened in Paris and Hamburg.

Fishes keeping in an aquarium tanks, though a fairly costly hobby, can be profitable when compared to other hobbies. The movements of colourful fishes in an aquarium would certainly please the ailing and convalescing people. The aquarium may gladden one's heart and broaden one's mind.

Aquarium Size

The size of the tank would depend on the number, size and type of fish to be maintained. Different types of tanks such as rectangular angle-iron tanks with glass sides and glass bottom reinforced with metal and single-piece acrylic tanks are available for use. Generally the shallow and wide tanks are preferable, as they have more surface area to facilitate oxygenation through atmosphere contact. Further such tanks can accommodate more number of fish than tanks having lesser surface area but with equal quantity has to be selected. In order to get a good view of both of fish and of plants in an aquarium tank, the light for the latter should come obliquely. Generally a place nearer to window side may be selected. The support or table positioning the tank must be strong and inn level. The following sizes of tanks are normally employed for keeping aquarium fish:

Table 25.1: Standard Aquarium Sizes

Length (cm)	Breadth (cm)	Height (cm)
45	25	25
60	30	30
75	30	30
90	30	40

If, the tanks are purchased, they should be checked for leakage; their sides should be cleaned with one per cent potassium permanganate solution, besides repeated washing with tap water.

In next phase of setting up an aquarium is the selection of small stones and fine gravel for spreading at the bottom of tank. For aquarium without biological filter, arrangement of smaller stones should be made in groups at the bottom and after that the gaps should be filled with fine gravel. Then the gravel bed is covered with a layer of coarse river sand. The ground layer should be 3-5 cm thick and should slope slightly towards the front of the aquarium. For the aquarium tanks where biological filters are to be set up, 3 to 6 mm sized dark particles of granite, basalt, quartz or slate should be ideal as bottom materials to help filtration. The calcareous rocks and coral stones should not be used as they may affect the water quality by altering the pH. The particles should be washed in tap water several times before they are used.

Filling Water

Rainwater free from contaminates or tap water is ideal for aquarium tanks. As tap water of domestic supply is likely to have good amount of chlorine which is harmful to fish, the same may be stagnated for about one or two days followed by subjecting it to strong aeration before use. The water must be filled in tank without disturbing the bottom sand. This may be done by means of a hose pipe running first into a cup kept on the aquarium bottom. The overflowing water from this cup will gently fill the aquarium tank without disturbing the bottom. Alternately, water may be gently filled on a newspaper spread over the bottom. In this way the aquarium tank can be first filled to about 5 cm and then filled directly.

Gravel

The gravel in the tank provides support for plants, a means of filtration with under gravel filter systems, and a region for fish to carry out activities such as breeding and feeding. The composition of the gravel is important in that gravel that contains minerals will dissolve and harden the water. If plants are kept in the tank, the gravel should be 2-5 mm in diameter. The gravel can be added to the tank and arranged levelly or terraced. To terrace the gravel, use flat rocks, wood or glass strips. Before adding the gravel to the tank, the gravel should be carefully washed to remove all small particles.

Planting

Plants are an important prerequisite of an aquarium. During the daytime, plants use the carbon dioxide, produced by fish and water to produce oxygen and energy. The oxygen is used by fish for respiration. At night, there is no sunlight for the plants to carryout photosynthesis, so the plants must rely on respiration to make energy. So, plants take in oxygen and produce carbon dioxide. Because of night time plant respiration, the carbon dioxide level in an aquarium rises at night, but once the light is turned on, the carbon dioxide levels drop due to plant photosynthesis. Each plant has to be fixed firmly deep into the bed of the sand. No green part should be under the sand. After planting, the aquarium is left undisturbed for a week, before introducing the fishes. Plants provide shade, shelter and sanctuary for aquarium fishes. Besides decorating the aquarium tanks they also serve in spawning and as food for certain fish. Before planting, the plants must be well washed; otherwise the disease causing parasites may attack the fish at a later stage.

The common aquarium plants are listed in Table 25.2.

Depending upon their size and shape, the aquarium plants are of different types *i.e.*,

Foreground Plants

These are small, low growing that often form carpet like matting by producing numerous runner plants. Foreground plants often inhabit shallow water and may require bright lighting. These are planted in front of middle ground and background plants.

Table 25.2: List of Common Aquarium Plants

Aponogeton crispus, Aponogeton madagascariensis, Echinodorus paniculatus, Echinodorus bleheri, Echinodorus cordifolis, Elodea canadensis, Hygrophila difformis, Hygrophila salicifolia, Hygrophila polysperma, Hygrophila stricta, Lemna minor, Myriophyllum mattogross, Myriophyllum spicatum, Marsilea drummondii, Nymphaea lotus, Nymphoides aquatica, Sagittaria graminea, Sagittaria subulata var subulata, Vallisneria Americana, Vallisneria gigantean, Vallisneria spiralis, Glossostigma elantinoides, Limnophila aquatica, Limnophila sessiliflora

Middle Ground Plants

Middle ground plants are medium sized which can be used behind foreground plants, infront of background plants.

Background Plants

Background plants are usually tall. These are fast growing plants, which require less light than foreground and middle ground plants.

Bunch Plants

Bunch plants are usually middle ground or background species that look good in groups of several. Bunch plants are often easily propagated by cuttings.

Specimen Plants

These are large plants. Specimen plants are often used as a focal point and may be highlighted with a spot light.

Contrast Plants

Different-looking plants can be used as a contrast to the other plants in the aquarium. Red leafed plants can be used as a colour contrast to green plants, while plants with pointed leaves can be used as a shape contrast to those with large round leaves. When contrasting plants, plane plants with similarities in colour, size or shape away from one another, while planting plants with differences closer together.

Floating Plants

Floating plants require plenty of light, but must protected from leaf burn by leaving distance between them and the bulb. Floating plants often propagate very quickly by division.

After planting is complete, the water filling is continued and maximum water level may be around 5 cm below the top of the tank. Fish may be introduced two days after the setting up of the tank.

Wood

Wood provides a refuge, a spawning site, and nourishment for some fish. Wood can further add to the acidity of the water, benefiting fish that prefer acidic water. Don't use wood in tank with fish that require hard, alkaline water, as the wood will affect the alkalinity.

Equipments Used for Maintenance of Aquarium

Heater

Generally heaters are used in cool places. The glass immersion heater is very commonly used in aquarium. These are of two types *i.e.*, non submersible and submersible. The submersible heaters are a beeter investment, because they are usually more reliable and need not be unplugged whenever the water drops more than 15 cm from the top. The heater must be unplugged for 10 minutes before it leaves the water. If the heater is immersed, it is subject to breakage. These types of heaters are generally fairly inexpensive. Selection of right size of heater is essential. Generally 2-3 watts per gallon of water is recommended.

When working in the tank, always unplug the heater for safety reasons. Some fish species may rest or hide on the heater. The fish often receive burns. To prevent this problem, protect the fish by surrounding the heater with a mesh, cage like structure.

Thermometer

It is used to monitor the aquarium's temperature. Several types of thermometers are available including stick-on liquid crystal types, floating glass types, and digital types. The liquid crystal type is convenient in that it is easily read when affixed to the side of the tank, but is not entirely accurate because temperatures outside of the tank can influence it. Glass type thermometers are not accurate but will serve the need of most aquarists. The most accurate, easiest to use, but more expensive of the three types is the digital thermometers, which gives a reading every few seconds.

Air Pump

The air pump can be used to power air stones which drive under gravel filters, internal box filters, and sponge filters. Separate air stone can be used for further aeration. The major drawback to air pumps is the noise they produce, especially when they are vibrating against something. Less expensive models are often noisier than higher quality, more expensive models.

Filters

Filtration is used to maintain good water quality. The filtration is of three types *i.e.*, mechanical, chemical and biological. In mechanical filtration dirty water carrying suspended waste particles and debris is passed through a medium, which physically traps the solids and allows the clean water through. The filter medium is usually nylon floss, filter wool or aquarium foam. Chemical filtration employs a highly porous surface to absorb some of the soluble waste products, such as NH_3. Charcoal and resins are suitable media for chemical filtration. In biological filtration, water is passed through a medium coated with bacteria that actually feed upon excretory products rendering them harmless in the process. Some suitable media for the bacteria to grow are the aquatic gravel itself, extensively perforated foam or recently, sedimentary rock chips have been incorporated into a trinkle feed system. The commonly used filters in aquaria are boxfilters, spongel foam filters, undegravel filters, reverse flow filters and power filters.

Bucket

A bucket is needed for water changes and adding water. A 8-10 litre bucket is sufficient. The separate bucket should be strictly used for the aquarium.

Siphon Hose

A siphon hose is needed for water changes. Siphon hoses are available in a range of sizes and designs.

Net

Sometimes it becomes essential to catch a fish from the aquarium and transfer it to elsewhere. At this time, a handier rectangular net is required. With the help of net, fish may be netted in the corner of the aquarium and it cannot make a dash for freedom during which might be injured. Every aquarist should have at least one net. The net should be fine mesh designed for aquarium use. Hand nets are generally made of mosquito nets or bolting silk.

Glass Cleaner

To clean the glass tank, fine steel wool is held in the figures by some fish keepers. An ordinary razer blade is also clipped at the end of a wooden handle to serve the same purpose.

Light

One of the most important ingredients to a successful aquarium is strong lighting. Light besides beautifying the tank helps in the photosynthesis of aquarium plants. Fish also require light to trace their food. Further light, which has a stimulating effect, is also known to influence the growth of fish. It is also reported that fish require vitamin D, an antibiotic from the direct sunlight to build up resistance to diseases. Strong sunlight destroys bacteria and keeps the tank healthy. For moderate tanks two bulbs of 60 watts each may be lit for eight hours a day. However it is better to use fluorescent lighting for promotion of plant growth and for even distribution of the light.

Introduction of Fish in Aquarium Tank

The number of fish for stocking depends mainly on surface area of tank, its dissolved oxygen content and size of the fish. However it is reported that 1 cm long fish may need about 75 cm^2 of the surface area. Based on the above, a tank of 75x30 cm size, for example may hold three fish each of 10 cm.

Healthy fish should be first quarantined for some time. Then are acclimated by mixing aquarium water gradually with water in which they are brought. It is also desirable not to feed the fish on first day. Fishes are generally introduced 2 to 3 days after planting when water would also be crystal, clear and oxygenated by the plants. Before introduction into the aquarium tank the fish may be treated with 2 per cent potassium permanganate solution to avoid parasitic attack. Aquarium tank is termed community tank when it is stocked with fishes of different species. However while stocking are should be taken to introduce compatible species, otherwise death or damage may occur.

Ornamental Fishes

Carassius auratus (Gold Fish)

The goldfish is the most common aquarium fish. It can be found in a number of varieties and colours. Mostly they are golden yellow, however the colour varies from grayish silver to all black. In natural habit, it reaches up to 45 cms. Goldfish is non-aggressive and very hardy fish. Careful maintenance is needed for good health. They need to be in the tanks that are not crowded and gives them sufficient swimming room. It is omnivorous in feeding habit. It feeds on plant matter, detritus, live feed, and dried feed. Goldfishes sexually mature within 5-6 months. It is prolific breeder. The males develop breeding tubercles during spawning time. Eggs are adhesive in nature.

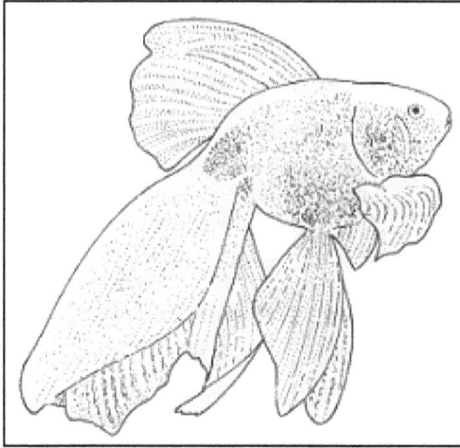

Figure 25.1: *Carassius auratus*

Pterophyllum scalare (Angel Fish)

Its origin is Amazon region of South America. It is peaceful but gets aggressive when spawning and guarding fry. The body is deep and flattened vertically. The dorsal, pectoral, pelvic, anal and caudal fins are large with elongated rays and erect in shape. The colour of angelfish is varying from grey to black. The angelfishes are commonly provided with 4-7 vertical colour bands among which the first band is curving through the eye. Males have pointed genital opening and that of the female is rounded. Eggs are laid on he water plants. Male as well as female take care of the eggs and youngones.

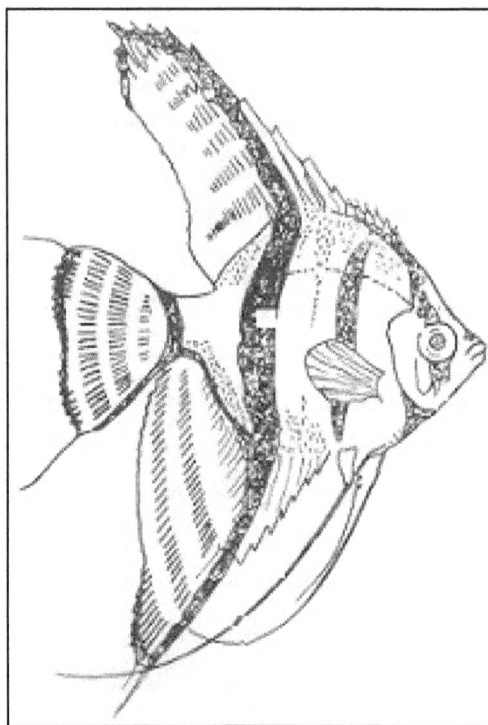

Figure 25.2: *Pterophyllum scalare*

It is an omnivorous fish. Its life span is 10 years. Among the most beautiful freshwater tropical fish, the angelfish comes in many varieties and attractive colours. They grow to large size and live for years. They comprise of two general groups. One group consists of fishes with different finnage, of which there are two types, normal-finned and long finned. The basic colours are black and silver. In young condition, they are suitable in a community tank, however they need a large tank as they grow in size. It does not disturb aquarium plants. Angelfishes are sold when active small.

Colisa lalia (Dwarf Gourami or Powder Blue Gourami)

Its origin is Ganges, Jamuna and Brahmaputra. It is very calm and harmless. Body is oval in shape. The dorsal and ventral fins are extended up to caudal fin. Caudal fin rounded to truncate and scales are large. The pelvic fin is modified into long thread like structures.

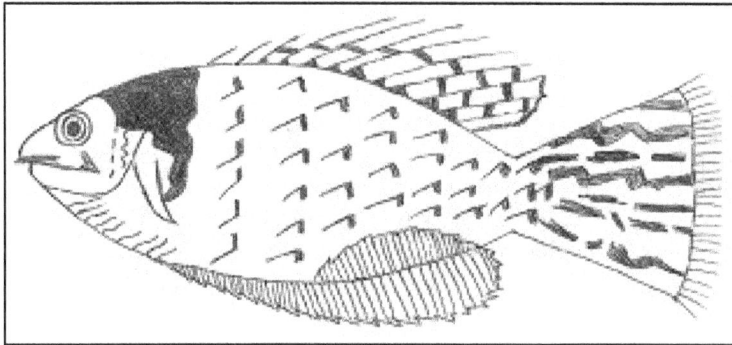

Figure 25.3: *Colisa lalia*

Background colour is orange. The anterior region is of blue green. The sides are with blue stripes and fins are red. Size of fish is about 6 cm. It is an omnivorous fish. In aquarium tanks it feeds on dry and live foods. Its life span is 3 to 4 years. Due to their peaceful nature, it is most suitable fish for small aquariums as well as community aquariums. They should not be kept with very large or aggressive fish. Diffused lighting in aquarium can show the colours of the males better. The males are dark coloured. These fishes build nest made of bubbles and plant materials. The male fish guards the eggs till hatching.

These lovely small sized fishes are often sold alive in bottles of water at Kolkata and they thrive well in an aquarium. It attains 5 cm length.

Betta splendens (Siamese Fighting Fish)

It is native of Cambodia and Thailand. The body is slender and dorsal fin is high. The caudal fin is long and broad so also the anal fin. The pelvic fins are pointed and small. The colour of fish is blue with red patches anteriorly. It grows up to 6 cm. Male fighting fishes are aggressive. Sometimes it jumps out of water; hence the tank should be provided suitable cover. The males should not be put together. They fight until one of them dies. It feeds mostly on live foods, flakes food and daphnia. It sexually mature within 4-5 months and builds bubble nests among water plants. The males are known for parental care.

Poecilia (Lebistes) reticulatus (Guppy)

The fish has been named after an aquarist, Rober John Lechmere Guppy who had sent a few living pairs of this species to the British Museum, London in 1860. As guppies are collected in large numbers from freshwater bodies, they also bear the name 'million fish'. Though the guppies had their origin in West Indies, they are grown in many countries as ornamental and larvicidal fish. It was introduced in India from South America in 1908. In India, it is now well established in certain parts of Tamil Nadu, Andhra Pradesh, West Bengal, Orissa and Maharashtra.

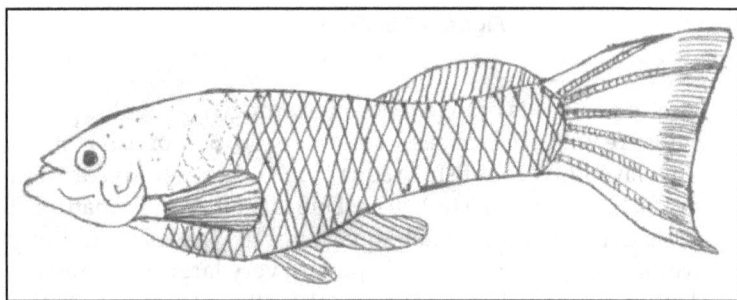

Figure 25.4: *Poecilia reticulata*

Body cylindrical with moderate mouth. Small teeth present on jaws. The dorsal fin is irregular and the eyes are large. Mouth is upturned. The common colours are red, green, blue, grey and yellow with spots. Young males are easily distinguished from the females by their beautiful orange, red and black dots all over the body and fins. The females are olivaceous. Males are always smaller than females. Gravid females can easily be recognized by the young they contain and the adult males are marked by the elongated anal fin which is used as intromittant organ by the male to guide spermatozoa into the females. Males grow up to 3 cm and females up to 5cm. It is omnivorous and voracious feeder. Female releases 15-25 youngones at a time. The optimum water temperature range for breeding is 20-24°C.

Etroplus suratensis (Banded Pearlspot or Stripped Chromide)

Body is deep, short, oval and strongly compressed. Eyes large, its diameter 3 to 4 times in length of head. Mouth small, teeth villiform, in a single row anteriorly but in one or two rows posteriorly on both jaws. Caudal fin slightly emarginated. Scales weakly ctenoid. Colour is light green with six to eight not very prominent vertical bands. Most of the scales above lateral line with a central white pearly sopt. Some irregular black spots on abdomen are present. Dorsal, caudal, pelvic and anal fins bluish or dirty green in colour. Pectoral fins yellowish with a black blotch at its base. Fish become sexually mature in second year. Females guard the fertilized eggs. It is a delicious food fish, especially when large. Young fish are very attractive for aquaria. It grows up to 25 cm.

Heleostoma temmincki (Kissing Gourami)

The body colour is white or pale yellow and devoid of pigment. Fish shows unique behaviour of kissing each other. The individuals of opposite sex extend the thick fleshy lips and kiss the other individuals. They are widely distributed in Southeastern Asian countries. It matures at 12 inches in size.

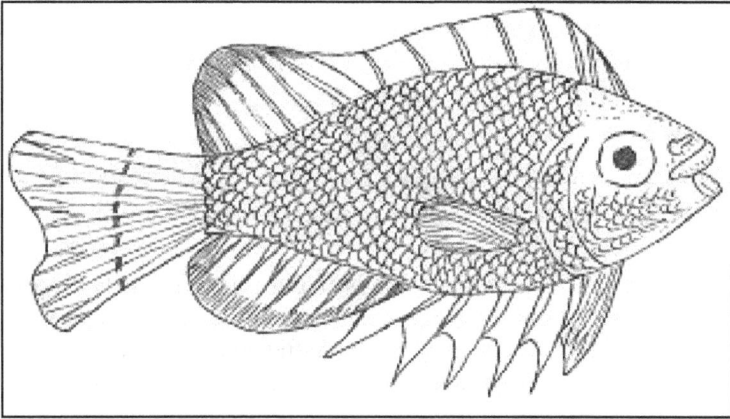

Figure 25.5: *Heleostoma temmincki*

Xiphophorus helleri (Red Sword Tail)

It is a small fish, with elongated body. Its eyes are large and fins are small. The mouth is slightly upturned. The lower rays of the caudal fin are modified into a sword like extension in male. The sword is not a weapon but for attraction. The number of young released by female being 20-200. It grows to 8 cm in length without sword. It lives in small schools and can be kept with other livebearers.

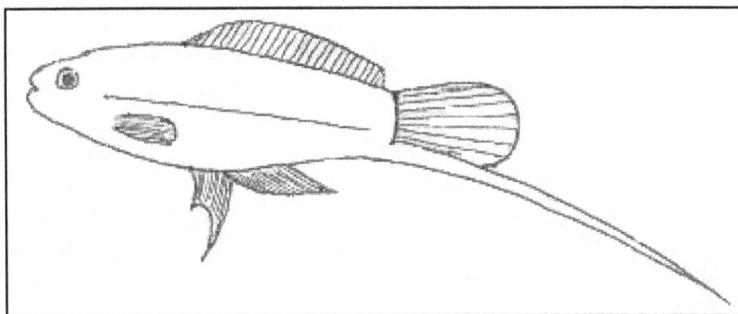

Figure 25.6: *Xiphophorus helleri* **(Male)**

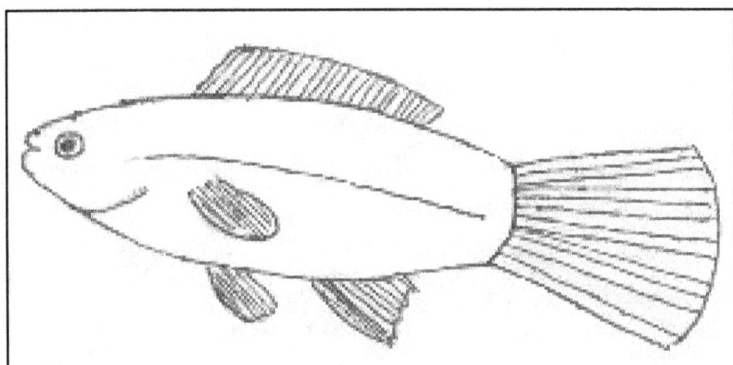

Figure 25.7: *Xiphophorus helleri* **(Female)**

The females are livebearers. It feeds on live food and easily accepts pelleted feed. Males are smaller in size than females. The parents may eat their youngones. Hence parents and younones are separated after breeding. Optimum water temperature for breeding is 23°C.

Danio malabaricus (Malbar Danio or Giant Danio)

It has relatively narrow distribution confined to South India, Sri Lanka and Burma. The body is silvery and tinted with several horizontal wide stripes of gold and blue colour. It has compressed body with round abdomen. Easily acclimatized to the captive conditions, hardy, attractive and quite peaceful. It is an ideal fish for community aquariums. Although extremely active, its movement is very graceful. It is omnivorous and can grow up to 15cm. The fish always remains towards the upper layers of water column and is a compatible species in community tanks. There is no sexual diamorphism except the soft, bulged belly during the ripe condition of the female.

Rasbora daniconius (Rasbora)

It is a beautiful and hardy fish. A bluish black lateral stripe along the center of the slim body that extends from the tip of the snout to the tail, adds to its beauty. It gets easily acclimatized to the captive conditions. It is very compatible, gregarious and quite peaceful.

Puntius ticto (Ticto Barb)

Ticto barb is a deep bodied small fish. Barbels absent and eyes are golden. It has shining silver belly. It has a black spot on the side of the tail before the base of the caudal fin and immediately behind the anal and a smaller one at the commencement of the lateral line. The dorsal fin of the male is reddish. It is prized for its iridescence and the red edging on its dorsal fin, which takes a deep ruby hue during mating time. It attains a length of 100 mm.

It is very compatible, active and peaceful which makes it suitable for community aquaria. It thrives well in a well-lit tank with plants having delicate leaves.

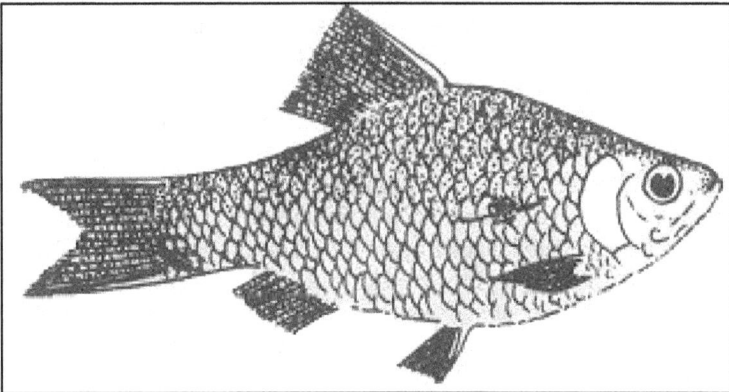

Figure 25.8: *Puntius ticto*

Chanda nama (Elongate Glassperchlet)

It is a small ovate fish with a fairly transparent silvery yellowish body, large mouth, black eyes and with fins bright orange. Upper half of the dorsal with deep black, caudal dusky and orange with a pale outer border. These small fishes are also used as for food in Uttar Pradesh, Bihar and West Bengal. Being a small fish with transparent body and coloured fins; it is a good species for aquaria.

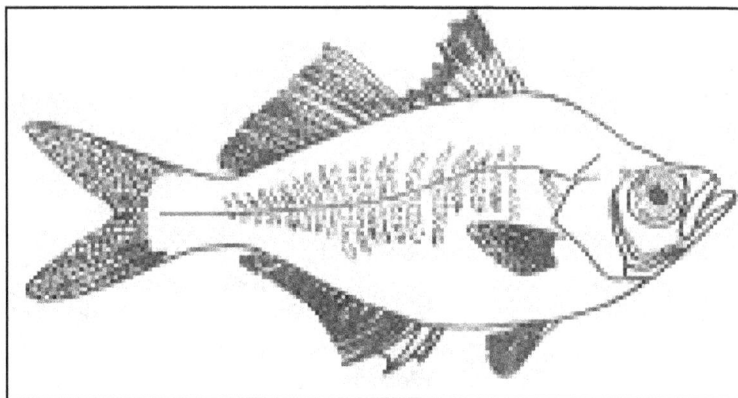

Figure 25.9: *Chanda nama*

Botia lohachata (Pakistani Loach or 'Y' Loach)

Body oblong, short and moderately deep. Abdomen is rounded. Dorsal profile is slightly more convex than the abdominal one. Snout conical and ventrally flattened. Mouth small, subinferior and narrow. Lips thick and fleshy, lower lip is feebly fringed. Jaws equal without any teeth. Four pairs of barbels present. Lateral line present and complete. Scales present but very indistinct. Scales on head absent. Dorsal fin arises infront of the ventral. Anal fin short. Caudal fin is deeply forked. Body colour is yellow. There are numerous black bands arranged variously on the body mostly 'Y' shaped. There is a median longitudinal black line over the snout, a band with a dorsal loop in the occipital region., three loops on the dorsum of the body, one anterior, one posterior and one over the dorsal fin. There is a ring like band at the base of the tail. It attains about four inches in length. It is found in Indian rivers.

Esomus danricus (Flying Barb)

Body is elongated and strongly compressed with more or less straight dorsal profile. Abdomen is rounded. Mouth opening is obliquely directed upwards and is small. Lips are thin and lower

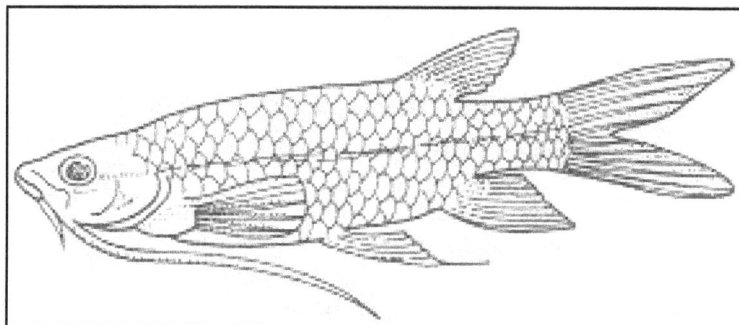

Figure 25.10: *Esomus danricus*

jaw is prominent. It has greenish body with a steel blue band in the flanks and a black streak along the dorsal surface. The maxillary barbel extends up to the middle of the pectoral fin. Being very delicate, they lose scales and may die immediately after catching, if special care is not taken. Caudal fin forked and eyes are placed inferiorly, visible from below ventral surface. They always remain in the surface layers of the water column. It is compatible, peaceful and fast swimmer, always moving from one end of the aquarium to the other. They prefer live feed. Its eggs are partially adhesive. It attains about 5 inches in length.

Mollienisia latipinna (Broad-Finned Mollay)

Male is slightly larger then female. It is magnificent with dorsal fin. Body is green with yellowish tinge. It is an omnivore. Male is

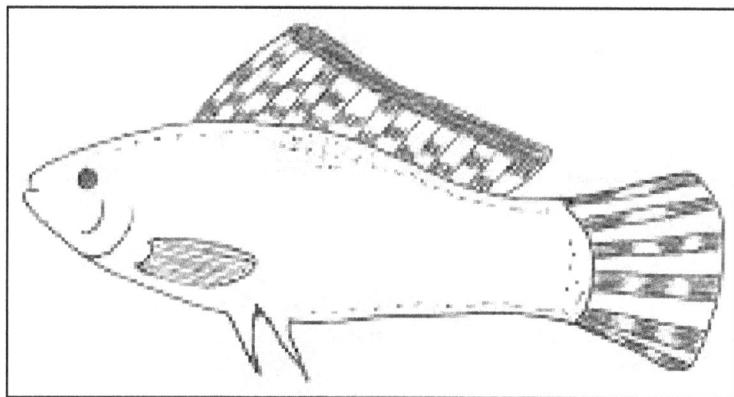

Figure 25.11: *Mollienisia latipinna*

with gonopodouim. The number of young released at a time by female being 30-70.

Barbus conchonius (Rosy Barb or Red Barb)

Body deep and compressed. Mouth moderate and barbels absent. The flanks are bright orange to red and metallic. Black spot is present near the caudal peducle. Lateral line incomplete. Colour is back shining olive-green; flanks and belly silvery tinged with reddish, shining ink red at spawning time. In spawning season the male changes to a lively iridescent blushing pink. The fish is docile and can be kept together with other ornamental fishes.

Feeding

Algae, which often grow around stones and in water, serve as a good food source to swordtails and kissing gourami. The fish should be fed lightly every day. Live protein rich foods such as cyclops, daphnia, tubifex, artemia nauplii, chironomid larvae, mosquito larvae etc are considered excellent. Fresh foods such as chopped earthworm, shrimp, fish paste, scraped boiled fish or raw liver also constitute good items.

Artificial pelleted fish feeds meant only for aquarium fish are easily available from shops. Dried prawns are powdered to which white and yellow of egg are added and the entire mixture is made into a paste with little salt. This is then pelleted trough a hand pelletiser and dried in shade. Such pelleted feeds may be used whenever required.

A feed-ring is ring shaped for confining dry feeds within the ring, avoiding its spread all over surface of water; it also helps in allowing the food, to fall at the bottom in only one place. Feeding ring is normally made up of plastic to float over the water surface. It is always better to feed the fish with different kinds of feeds. Fish may be fed once or twice a day according to their preference.

Cleaning of Aquarium

Excess food and detritus may be removed 30 minutes after feeding by siphoning out with rubber tubing. If the sides of tank are found with algal deposition, the same may be scraped by an aquarium scraper, which is a long handle with a clip at one end for fixing a razor blade. In case of the tanks with biological filter, it is necessary to rake the gravel bed periodically to liberate the unwanted gases and ensure a continued water flow. If an aquarium is neglected for a long period of time, its inhabitants suffer. Algae may build up on the glass and rocks if no algae eating fish is kept.

Chapter 26
Freshwater Pearl Culture

Scenario of Pearl Culture

Pearl has been adored across the world as a precious gem from time immemorial. In 1600 AD certain scientific information regarding the origin of pearl in oysters was available. By the beginning of the present century Herdman (1903-1906) carried out extensive scientific investigations on the formation of pearl within the bivalves. It was Tokichi Nishikawa who was the first to offer scientific explanation as to the origin of pearl which earned global acceptance. The hypothesis he propounded is known as 'pearl sac theory'.

The world trade of pearls which is over US$ 2 billion per year is controlled by a few countries. Pearl culture is by now a good commercial venture in some of the advanced countries. In nature pearl is formed in mussels by two distinct ways-one through accidental intrusion of a foreign body into the mussel and the other by the occurrence of a wound within the muscle by any chance. But such natural pearls are scarcely available as the incidence of the above two phenomena is also rare.

Pearl obtained through scientific culture is popularly known as 'cultured pearl'. On the other hand 'imitation pearls' are those, which are derived from the plastic or glass beds by addition of artificial luster. This imitation pearls although they look like real pearls are in great supply in the open market and are very cheap.

It is perhaps China that introduced pearl culture in the frsh water lake of the middle China in 1200 AD. The Chinese practiced a method in which the mussels were produced in bamboo cages and pearls were cultured through implantation of a foreign particle of clay or a minute idol of the lord Buddha inside the mussel. With the passage of time through many changers pearl culture became an established practice and from 1960 onwards-rapid progress could be sighted.

Freshwater mussel (*Lamellidens marginalis*) is available in plenty In India, Bangladesh, Burma, and Sri Lanka. This type of mussel is generally found in ponds, lakes and rivers. They live in holes, which they make with the help of their big legs. The mussels chiefly feed on planktons and attain maturity at an age of two years. They breed almost throughout the year particularly during the rainy season.

The Central Institute of Freshwater Aquaculture (CIFA), Bhubaneswar started research in 1987 on freshwater pearl culture and evolved a base technology for growing pearls in freshwater.

Pearl formation is an abnormal response in the normal biological processes of freshwater mussels. The mantle tissue is the important organ responsible for producing layers of shell including the pearl nacre on the interior of the shells. A pearl has more or less similar structural and chemical properties to that of the mother of pearl layer of the shell interior. The outer side of the mantle tissue is the 'key note' in the 'orchestra' of pearl formation. When an external stimulus such as accidental trapping of a hard foreign body or a parasite infection or a lesion take place in between the layers of the mantle, it leads to deposition of micro-layers of pearl nacre around the source of stimulus, resulting in a natural pearl. The secretary function of the live mantle forms the basis for modern pearl culture.

When pearl is to be produced artificially, two different methods are adopted. In the first method a piece of graft tissue is introjected into another mussel. In the second method a hard ball-like material called bead along with flesh is pushed into its body. In winter as the secretion is less, thus season is not suitable for pearl culture. The ideal temperature of water for this purpose is 15°C to 30°C. As the mussels become weak during breeding season; this period may also be avoided for culture.

For artificial production of pearl a fully-grown healthy mussel is required. The mussels must be washed clean and immersed in a pot of water at least one day before the introjection. The next day the mussels will have to be washed afresh. From among these a big healthy mussel is chosen which is known as a 'donor'. A thin and long portion from the side of its shell (mantle epithelial layer) is cut out. After cleaning it this piece is placed on a glass slide and cut further into smaller pieces and 2-4 drops of water –soluble Eosin is added on them. These pieces are then kept under a cover. This process helps to keep the pieces of muscle fresh. The other mussels are placed in tub full of water in such a fashion that the shell valves are placed upwardly. By this technique, valves of the shell can open easily, otherwise it becomes a little difficult to open them. A pinch of menthol may be sprinkled over water for easy opening of valves. Then as a mussel starts to open its valves a piece of wood is inserted with the help of valve opener and the mussel is finally placed on a mussel-holder stand. The pieces of graft tissue are then introduced into the mantle epithelial layer. One to five pieces of this type of graft tissue may be introduced if the mussel is a big sized one. After the operation the mussel is removed from the stand and put into water. The introjected mussels are washed thoroughly and kept in a tub of water for over night. Next day they are again washed before being dropped into a pond.

According to the second method, a small hard round object, made from conch shell, is inserted along with the piece of graft tissue. At present small beads made from conch shell are used for this purpose. Along with the graft tissue this type of beads may be introduced in the mantle epithelial layer of a mussel. Dorsal side of graft tissue should touch the bead. This graft tissue makes a sac within a short period. After that the nacre secretion starts depositing on it leading to the birth of a pearl. The first method takes about 18 to 24 months. But in the second method only a year is required.

The culture of pearls can be done along with fish culture. It is therefore, expected that pearl culture will be an economic proposition in near furutre. A fish-cum-pearl culture pond may be considered ideal, where the fishes and pearl oysters can grow well and their natural food is available in plenty.

Fish as well as mussel is in a state of equilibrium with the environment and the pathogens. A change in the environment disturbs this resulting in a stress response in the aquatic organisms. The most extreme stress response is mortality, but below this level there may be several other responses *viz.* 1) changes in their behaviour 2) reduced growth/food conversion efficiency 3) reduced tolerance to disease 4) reduced ability to tolerate further stress. The environmental parameters to be monitored in fish-cum-mussel culture are:

Dissolved Oxygen

Often fishes swim on the surface of water gulping air with mouth wide open. This stressed condition is due to oxygen depletion in pond. For mussel, a pumping mussel removes about 5 per cent of the oxygen content of the water passing through mantle cavity. During periods of mantle closure, which may be long, they build up oxygen debt, which has to be rapid when they are able to restart pumping. Moreover, since pearl culture is mostly done in stagnant water dissolved oxygen concentration should be maintained above 3 ppm. This is maintained by aeration.

Ammonia

It is commonly the second important parameter after dissolved oxygen. The safe concentration of the uniodised ammonia (gaseous form) in water for fish and mussel culture is 0.02 to 0.05 mg/liter.

Hydrogen Sulphide

Very often the pond muck in the culture ponds smells like rotten eggs. This is due to accumulation of hydrogen sulphide gas, which is produced by chemical reduction of organic matter. Such condition is not congenial for fish and mussel culture. The muck has to be removed either before starting the culture operation or during operation, raking should be done along with liming, at the rate of 50 kg/ha during daytime.

Suspended Solids

They originate from phytoplankton blooms, uneaten food particles and fish faecal matter. Suspended solids are important in reducing the penetration of light thus reducing productivity. For fish cum mussel culture a transparency of 30 cm is ideal.

pH

It is an important parameter affecting fish and mussel health. The optimum pH for both fish and mussel is between 7 and 8. It is because neutral or weak alkaline water is good for pearl culture.

Colour of Water

Generally considering the planktonic food requirement of both fish and mussel, a yellowish green colour of pond water is taken as ideal. It indicates that diatoms and green algae, which have brown and green pigments respectively, are present and influence the colour of water.

Nutrient Content of Pond Water

Since the algal flagellate population constitutes the important feed of mussel sufficient quantity of nitrate and phosphate is essential for stimulating their growth. Moreover, since calcium is the main constituent of mussel shell and pearls an optimum level of 16 mg/liter should be maintained.

Diseases of Mussel

It is important to mention that since mussel culture has been very recently initiated in India there is practically very little literature on mussel disease. However some common diseases those are likely to be encountered during culture and the mechanism of the defense of mussel is mentioned.

Bacterial Pathogens

Mussels live in an environment rich in bacteria *i.e.*, in the pond and their filter feeding results in ingestion of many kinds of micro-organisms including pseudomonas which are potentially pathogenic.

During implantation for pearl culture operation of the mussel is done and cut pieces are implanted. Here there is every possibility of infection by bacteria as such antibiotics are used to heal up the wound.

Fungal Pathogens

These may be important as invaders of secreted structures of shell in mussels. Most of them are aquatic phycomycetes.

Mechanism of Defense in Mussel

Different four kinds of defenses exist *viz.* phagocytosis, encapsulation, hemocytosis and nacrezation.

Phagocytosis

By this mechanism foreign material (including pathogens) is taken up by haemocytes and removed from contact with tissues. Several parasites and pathogens of molluscs have been reported as being phagocytized.

Encapsulation

Any invasion by a parasite or other foreign body too large to be phagocytized is encapsulated. Here concentric layers of cells are formed around the invader.

Hemocytosis

Any parasitization or microbial infection of mussel is followed by increase in the number of haemocytes.

Nacrezation

It is a defense mechanism of mussel in which nacre secreted by the mantle epithelium is deposited around parasites or foreign bodies, which invade and irritate the mantle region. Pearl formation is an outcome of nacrezation. Induction may be by invasion of certain trematode metacercaria, by sand grains, or by artificial implants of mantle epithelium in subepithelial tissues as is being done in pearl culture.

Chapter 27
Sex Manipulation Techniques

During the last two decades, there has been the emergence of certain advanced technologies in fisheries sector. In terms of achieving sustainable levels of production characterized by market-oriented growth levels, however, the farming systems aforesaid suffer from limitations. One approach identified by scientists to wriggle out of this constraint and to augment sustainable production is to resort to sex manipulation as a tool for the purpose. Changing of sex of fishes to be farmed in their early stages of life is known as sex manipulation.

Sex Reversal in Nature

In few fishes sex reversal takes place naturally. They are males during initial stage of their life and lateron, they get changed into females, and vice- versa.

Types of Sex Reversal

It is of two types, *i.e.,*

Protogynous Type of Sex Reversal

Fishes are male in their initial stage of life and in later life they get transformed into females. Such types of fishes are known as protogynous fishes. Groupers (genus Epinephalus) shows protogynous type sex reversal after a certain age.

Protandrous Type of Sex Reversal

Fish are females in their initial periods of life and in later period they get transformed into males. Such fishes are known as protoandrous fishes. Gilthead sea bream (Spams aurata) show protandrous type sex reversal from 2 to 4 years of its life.

Techniques of Sex Manipulation

This consists of 1) use of chemicals 2) chromosome/genome manipulation by way of gynogenesis, androgenesis and polyploidy and 3) selective breeding.

Use of Chemicals

Different chemicals are identified which are capable of changing sex among fishes. Among these, male sex inducers are 17, Alpha methyltestosterone, 19 norethissterone, 11 ketotestosterone, 17 alpha ethyl testosterone, testosterone propinonate, andro stenediol, methyl androstenediol etc. Of these 17 alphal methyl testosterone and 19 norethisterone are the most effective chemicals.

Different estrogen derivatives serve as female sex inducers. These have been used successfully to induce feminisation in various fish species. These are 17, betaoestradiol, oestrone, diethyl stilbestrol, and oestradiol butyryl acetate. Of these 17, beta estradiol has been found to be the most effective.

Time of treatment for sex reversal in fishes is specific. In the case of fishes, whose gonadal sex differentiation starts after hatching, treatment should be given just after hatching or before initiation of feeding by the immersion of eyed eggs and larvae in inducer solution. In viviparous fishes such as Guppy, the gonadal sex is differentiated just before parturition. Therefore, treatment should be started at the embryonic stage in ovary by feeding the gravid mother (female containing egg in its ovary) with hormone containing feed.

For use of hormone containing feed, steroids are dissolved in alcohol. This liquid is mixed in feed by spraying uniformly over it with the help of syringe. The alcohol soon evaporates and steroid gets absorbed in the feed. These feeds can be stored at 15 °C and the required amount of feed is used every day.

Chromosome/Genome Manipulation

This is done adopting gynogenesis, androgenesis and polyploidy.

Gynogenesis

For obtaining female individuals only, gynogenesis process is applied. It is the normal development of eggs derived from maternal genome and without paternal genetic inputs. In this process, eggs are activated by irradiated milt containing sperms. Immediately after fertilization, a cold/heat shock is given when the egg is at meiotic metaphase II stage. Due to shock, metaphase is arrested with the retention of polar body, resulting in the development of diploid (2n) gynogenetic offsprings. Thus exclusively female offspring could be obtained through this process.

Gynogenesis can also be used to produce the monosex population (all females with female homogamety). Like Indian Major Carps, where females weigh heavier than males of the same age group, the monosex population can also be produced in case of mahseer and other endangered fish species.

Natural gynogenesis is reported in some members of the family Poecilidae, *e.g.*, *Poecilla formosa* and cyprinidae, *e.g.*, *Carassius auratus gibelio* reproduce through natural gynogenesis.

For induction of gynogenesis first of all, matured male and female brooders are selected. Then ovaprim/ovatide/pituitary hormone injection has to be given. Milt may be stripped from males an hour before the expected ovulation time of the female fishes. This milt may be mixed with freshly prepared Hank's solution in the ratio of milt one part and Hank's four parts. This milt mix may then be transformed into a small pteri dish to make a thick column. Then pteri dish may be placed on an ice tray over the mechanical shaker in the uv chamber for irradiation. The uv lamp and the mechanical shaker are operated at a medium speed. Care should be taken to ensure that the milt in the petri dish is properly and continuously mixed with a shaker and a cotton plug is applied outside the petri dish to prevent spilling of milt into the ice tray. The milt irradiation process nay be continued for 17-20 minutes in the case of major carps and 35-40 minutes in the case of common carp for complete irradiation of milt. After completion of irradiation, milt has to be stored in the refrigerator at 4-5 0C. Eggs are to be collected in a basin through stripping of gravid female and the eggs so stripped have to be mixed together with irradiated milt for the fertilization after adding a little water. The irradiated milt activated eggs are now given thermal/pressure shock to restore diploidy nature.

Androgenesis

For obtaining male individuals only, androgenesis process is applied. The process of induction of androgenesis is relatively uncommon in nature and is also difficult to induce artificially. Natural androgenesis was reported to occur when female common carp is crossed with male grass carp. However, the percentage of incidence was very low. Androgenesis can be induced eliminating maternal genome in a similar way as in done for the elimination or denaturization of paternal genome in gynogenesis and activation such genetically denatured eggs by normal sperm followed by usual temperature shock or hydrostatic pressure. As has been experienced in many countries, androgenesis can help in the conservation of fish biodiversity in India also since it could be possible to reconstitute the endangered species like mahseer and hilsa from their cry preserved milt, if available in the gene bank.

Polyploidy

Like gynogenesis and androgenesis, polyploidy also occurs in nature and can be induced too. Natural polyploidy occurs when certain species of fish are crossed with other species of remotely related genus as in the case of grass carp x big head carp hybrids. The hybrids of this cross are triploids. In fishes, polyploidy occurs in common carp and trout. It is because of chromosomal translocation.

Polyploidy is induced in the same manner as diploid gynogenesis. However, unlike the later, the former is induced by subjecting the fertilized egg (by normal sperm) to the usual shock treatments. Triploidy is achieved by preventing the extrusion of second polar body, while tetraploidy is induced by blocking the first cleavage. Generally triploids are supposed to be sterile and tetraploids are fertile. Being sterile in nature, these fishes grow faster than the normal fishes and they have bigger size because the energy contributes only in growth rather than gamete genesis. Swarup (1959) made efforts to produce triplod fish for the first time in India. Like triploids, the tetraploids can also be produced by disruption of the first cleavage mitosis. The results suggest that tetraploids also grow faster more than the normal fishes.

Selective Breeding

It helps in the production of a high proportion of one sex, which can be of great use in aquaculture to augment production. Sterile fishes produced by selective breeding are of greater use in controlling reproduction where it is not desired.

In India, under a selective breeding programme, a variety named Jayanti rohu (CIFA-IR-1) has been successfully developed in recent years. This research was done under Indo-Norwegian collaboration during 1992-2000 at CIFA, Bhubaneshwar. The field testing of jayanti rohu has been carried out in Orissa, West Bengal, Andhra Pradesh and Punjab. A 17 per cent high growth realization per generation after four generations of selective breeding has been achieved.

Selective breeding is a long-term programme. It largely depends upon the life history especially breeding cycle of the target species. This method remains an effective tool to achieve genetic improvement in culture species. The technique may get a further boost through targeted selection, assisted with molecular markers to achieve genetic improvement in short time by integrating the traditional approach with modern biotechnology.

Adverse Effects of Sex Manipulation Techniques

Sex manipulations techniques may unleash certain adverse effects. These are (*i*) they may cause ecological imbalance, (*ii*) use of steroids may be fatal for humans, and (*iii*) the possibility of suppression of one partner either male of female over a period of time.

Chapter 28
Cryopreservation of Gametes and Embryos

Cryopreservation of sperm and embryo are important technological tools for ex situ conservation. The technique is useful in overcoming the problem of male fish maturing before female fish, and eliminating spatial and temporal barriers in breeding. It is also useful in selective breeding, hybridization and stock assessment. The technology of sperm cryopreservation has been successfully standardized for fish like Atlantic cod, rainbow trout, grey mullet, zebra fish, common carp and tilapia. The methods for cryopreservation of fish sperm have been developed for over 80 species of freshwater and saltwater fish (Rana 1995, Leung and Jaemison 1991, Figiel and Tiersch 1997). In India, the National Bureau of Fish Genetic Resources (NBFGR) has developed and standardized the technique for cryopreservation of fish milt. A mini gene bank with milt of Catla, Rohu, Mrigal, Common carp and Mahseer has been developed by NBFGR

Technique of Cryopreservsation

The biological material is preserved and stored at low temperatures, usually at –196°C, the temperature of liquid nitrogen. At this temperature the cellular viability can be maintained in a genetically stable form and is affected only by background radiation.

The technique involves cooling and freezing of aqueous solutions. The following steps of the technique are described.

Collection of Milt and Pre-Freezing Sperm Quality

During collection milt is usually contaminated by fish urine, mucous and water, which may affect the quality of cryopreserved spermatozoa. The duration of storage of milt prior to cryopreservation may also affect the post-thaw viability of the spermatozoa. Milt should be frozen immediately after collection.

Extenders

An extender is a simple solution consisting of organic and inorganic chemicals resembling that of seminal plasma or blood, in which the viability of spermatozoa can be maintained during in vitro storage. The chemical composition of the extenders used for cryopreservation of spermatozoa vary widely. However, simpler extenders, some containing only Nacl, $NaHCO_3$ and lecithin have been found useful.

Cryoprotection

To minimize the stress on cells during cooling and freezing, cryoprotectants are added to extenders. Generally ethylene glycol, dimethyl sulphoxide, dimethyl acetamid, propanediol and methanol are used as cyroprotectants for preservation of spermatozoa of bony fishes.

Equilibrium Period and Dilution Ratios

The optimum cryoprotectant concentration for a given protocol depends on the equilibrium period, the time allowed for cryoprotectant penetration into the cells. A equilibrium period does not appear to be necessary although preparatory procedure for cryopreservation usually takes up to 30 minutes before samples are cooled.

Various ratios of milt and diluent (diluent=extender + cryoprotectant) have been tested. In salmonids, similar results were obtained with dilution ratios (M : D) ranging from 1 : 1 to 1 : 9. For rainbow trout 1 : 3 was found as optimal dilution ratio.

Cooling and Thawing

These are among the most critical variables, which affect the success of cryopreservation. Many of the protocols, which relate to salmonids, use dry ice block as a coolant. It has been suggested that the optimum for salmonid spermatozoa may lie between 30 and 60°C/Minute. In tilapias, maximum post –thaw motility was achieved in samples cooled at 30°C/minute.

Storage

Diluted sperm samples have been successfully stored in polypropylene vials (1-2 ml) as pellets (40-200M1) and in 0.25 ml and 0.5 ml plastic straws. The polypropylene vials are available with various colours of *cap* inserts for easy identification. The vials may be stored in racks. Pellets of diluted semen are usually made by using a dry ice block (–79°C) as the coolant. This technique can be used in the field. Holes are drilled into a block of dry ice into which a fixed volume of diluted semen is added. After several minutes the frozen pellets are removed and stored in vials. Plastic straws are now more readily available and are also used for the cryopreservation of spermatozoa. Diluted seen is drawn into the colour-coded straws and either heat sealed or plugged with a special colour coded powder, which gels in the presence of a liquid to form a seal. The sealed and frozen straws are stored under liquid nitrogen. Liquid nitrogen (–196°C) is the most commonly used cryogen. Frozen samples are usually stored in the liquid nitrogen refrigerators and are either held in the vapour phase or immersed under liquid nitrogen. Samples should be labelled and colour-coded as far as possible for easy identification.

Cryopreservation of Embryo and Eggs

The cryopreservation of embryos and eggs has not been successful in any finfish species. It is due to permeation barriers in multicompartment system. Cryoprotectants are toxic to embryos at high concentrations and long exposure times. The cryopreservation of fish ova and embryos is quite complicated as compared to fish spermatozoa. The reasons are i) large egg size ii) the presence of two different egg membranes and the different water permeability of each of the membranes iii) higher level of degradation during cooling, and iv) presence of large volume of yolk.

Finfish ova and embryos are large. They contain a large amount of yolk and are covered with a relatively thick chorion. Uniformity in the penetration of conventional cryoprotectants and in cooling during the freezing process has not been attained. A few studies have been carried out to assess sensitivities of different development stages of embryos, towards various cryoprotectants and freezing parameters. Relatively, attempts to cryopreserve invertebrate larvae have been more successful than the finfish. In penaeid prawns, successful survival of thawed larvae has been reported up to freezing temperature –40°C. Research to develop cell lines, embryonic stem cells and germ cells, from Indian fishes to develop technology for cloning has been emphasized in the past. There have been some successful studied in developing cell structures such as ovarian tissue from immature ovary of *Clarius gariepinus* (Kumar *et al.*, 2001). Pluripotent cell line from sea bream embryonic stem-like cells (SBES1) has been reported from blastula-stage embryos of the cultured redsea bream, *Chrysophrys major*. Most recent significant efforts and success have reported in developing cell structures and cell lines from *To putitora, Labeo rohita* and *Lates calcarifer* in India (Lakra *et al.*, 2005, 2006 a, b). In future, development of expertise for other tools like embryonic stem cells preservation and cloning needs active consideration, to overcome the challenge of long-term storage of finfish eggs and embryos.

Recently, NBFGR and CIFA have invited research programmes on cryopreservation of blastomeres of Indian Major Carps and catfishes. The ultimate objective is to achieve germline chimeras, where the donor cells enters germline development and are able to give rise to fertilizable gametes. In most of the reports, pluripotent *blasomeres* when grafted into the recipient embryos have yielded somatic and occasionally germline chimeras, attributed to the relative low transmission rate of donor cells than the host embryo cells. Targeting the germ cells through precise identification and transplantation can improve the success of germline chimeras. Expression of primordial germ cells of donor into host system, with successful development of live trout fry has been demonstrated at experimental level. More research in this area may provide simple assays to target germ cells, facilitating pure line in vitro culture of primordial germ cells for grafting into host embryos.

References

Figiel, C. R. and T. R. Tiersch. 1997. Comprehensive literature review of fish sperm cryopreservation. Annual meeting of the World Aquaculture Society, Seattle, Washington. Pp 155.

Kumar, S., Bright Singh, I. S. and Philip, R. 2001. Deveopment of a cell culture system from the ovarian tissue of African catfish (*Clarias griepinus*). *Aquaculture*, 194: 51-62.

Lakra, W. S., M. R. Behera, N. Sivakumar, M. Goswami and R. R. Bhonde. 2005. Development of cell culture from liver and kidney of Indian major carp, *Labeo rohita* (Hamilton). *Indian J. Fish.*, 52(3): 373-376.

Lakra, W. S., N. Sivakumar, M. Goswami and R. R. Bhonde. 2006. Development of two cell culture systems from Asian Seabass *Lates calcarifer* (Bloch). *Aquaculture Research*, 37(1): 18-24.

Lakra, W. S., R. R. Bhonde, N. Sivakumar and S. Ayyappan. 2006. A new fibroblast like cell line from the fry of golden mahseer *Tor putitora* (Ham.). *Aquaculture*, 253: 238-243.

Lakra, W. S., K. K. Lal, Vindhya Mohindra. 2006. Genetic characterization and upgradation of fish species. *Fishing Chimes*. 26 (1): 256-258.

Leung L. K. P., Jamieson B. G. M. 1991. Live preservation of fish gamestes. In: Jamieson, B. G. M. (Ed.). Fish evolution and systematics: evidence from spermatozoa. Cambridge; Cambrige University Press. pp 245-269.

Rana, K. 1995. Preservation of gametes in Broodstock Management and Egg and Larval Quality. N. R. Bromage and R. J. Roberts, editors. University Press, Cambridge, England. Pp 53-75.

Chapter 29

Transgenic Fishes

The term 'transgenic' refers to 'an individual in which a transgene (an isolated gene sequence used to transform an organism) has been integrated into its genome'. Around 30 laboratories in about 10 Asian countries are actively engaged in transgenic fish research. Asian scientists are the first to initiate research in transgenic fish and Asia is the center of research activity in transgenic fish. Several techniques are currently available for transgenic fish production, which have been developed to increase the efficiency of transgene integration, or to produce a large number of transformed individuals simultaneously. Although these new methods of gene transfer are gaining importance due to the encouraging results reported (Tanaka and Kinoshita, 2001; Lu *et al.*, 20022; Grabher *et al.*, 2003., Hostetler *et al.*, 2003; Kinoshita *et al.*, 2003).

The first batch of transgenic fish was produced in China in 1984. This consisted of fast growing common carps. The transgene had a mouse promoter gene linked to human growth hormone gene (Zhu *et al*, 1985). Since then, scientific teams of USA genetically engineered carp and catfish for fast growth.

The most useful application of transgenic fish production technology is in stock improvement of commercially important fish species. The transgenic technology introduces genes encoding desirable traits into the genome of organisms in one generation, which is inherited by future generations resulting in a rapid development of new genetic stocks with desirable traits. Other applications of this technology are in providing a model system for basic research on gene structure, function and also for the production of specific proteins in fish.

Several studies have shown that growth enhanced transgenic fish have improved feed-conversion efficiency, resulting in economic and potential environmental benefits such as reduced feed waste and effluent from fish farms. Currently no transgenic animal has been approved for food production in the United States, although that may change.

A recent study of growth-enhanced transgenic and nontransgenic salmon found that transgenic salmon did not affect the growth nontransgenic cohorts when food availability was high. However, the survival of both transgenic and nontransgenic cohorts was deleteriously affected when feed resources are limited.

A company called 'Aqua Bounty' is currently awaiting regulatory review of its fish by the U. S. Food and Drug Administration (FDA). Aqua Bounty estimates that its Aqua Advantage salmon, a modified Atlantic salmon, reach commercial size in one-third of the time required for non-transgenic salmon (Fletcher *et al.*, 2001). The faster growth of each generation can lead to increased production per unit time along with savings on total food per pound of meat product (Aqua Bounty, 2002). The company uses a growth hormone and an antifreeze protein to increase the salmons' feeding efficiency and tolerance for cold waters. Private industry around the world is developing nearly 20 species of transgenic fish and shellfish, including catfish, carps, oysters and trout (FAO, 1999). To date, no country has approved any of these species for commercial production or human consumption. The FDA's consideration of the Aqua Advantage salmon is the first such case. Aqua Bounty has also applied for approval permission in Canada.

In India, research in transgenic fish was initiated at Madurai Kamaraj University (MKU), Center for Cellular and Molecular Biology (CCMB), Hyderabad and National Fathima Matha College, Kollam. The first Indian transgenic fish was generated in MKU in 1991 using borrowed constructs through Department of Biotechnology efforts. Experimental transgenics of Rohu, Zebra fish,

Catfish, and Singhi were developed and genes, promoters and vectors of indigenous origin made available for two species namely Rohu and Singhi for engineering growth. Using this, transgenic rohu produced from an indigenous construct at MKU has proved to be eight times larger than the control siblings. The transgenic rohu attains 46 to 49 grams body weight within 36 weeks of its birth (Ninawe, 2006). Jayanthi rohu, a selectively bred variety of fish was developed under NORAD (AVAFORSK)-CIFA project for selective breeding of rohu. Indian Council of Agricultural Research (ICAR) plans to develop autotransgenesis in commercially important fish species with growth hormone gene to avoid opposition of the society as an ethical issue. Production of pharmaceutical and other industrial products from piscine origin development of transgenic native glowfish varieties, fish biosensors for monitoring aquatic pollution isolation of genes, promoters and synthesis of effective gene constructs, researches in embryonic stem cells and *in vitro* embryo production are the future areas for promotion.

In year 2003, the production of transgenic ornamental fish (Glowfish) carrying fluorescent genes borrowed from jellyfish, has paved the way for producing new multi-coloured fluorescent fish. This development is significant from the Indian perspective because the western Ghats and the North-Eastern region are known to be rich in high value freshwater ornamental fish varieties. The significant use of transgenic fish would be increasing the reproduction capacity of economically –important animals, as also their potential to yield the desired product using transgenics in pharmaceutical research and production.

Tilapia has been extensively genetically modified and promoted as a transgenic fish exclusive for isolated or contained production. Transgenic tilapias, modified with pig-growth hormone, were three times larger than their non-transgenic siblings. Tilapia genetically modified with human insulin grew faster than non-transgenic siblings, and could also serve as a source of islet cells for transplantation to human subjects. Trout growth hormone was used to produce transgenic carp with improved dressing properties. Such transgenic carp are recommended for production in earthen ponds.

Transgenic fish that escapes into natural environment could be an environmental nuisance by becoming an invasive species. This danger mainly arises for those transgenic fish endowed with new genes that improve such fitness traits as breeding capabilities and the ability to withstand harsh environmental conditions.

Environmental groups in United States are ashing seafood retailers to pledge not to sell genetically engineered fish and to oppose their commercialization. Citing potential negative human health effects as a threat to the genetic purity of wild salmon, the environmental groups announced a new campaign on October 18, 2001 that aims to prevent the commercialization of genetically engineered fish (Anon, 2001). Transgenic fish of various species of Salmon, Tilapia, Channel catfish and others are being actively investigated worldwide as possible new food producing varieties. The groups cite a pending application to the U. S. food and Drug Administration (FDA) for market approval of an experimental salmon developed by Aqua Bounty farms. The company describes itself as a development-stage biotechnology company. The company developed the transgenic fish technology out of its research on fish species living in the high north that have a compound in their blood that lowers the freezing point of the whole fish so that they are safe from freezing through the frigid Arctic winters (Anon, 2001). The Food and Drug Administration could approve the application of any time, the groups fear, putting the first engineered fish on American dinner plates, grocery shelves and in restaurants across the country.

References

Anon, 2001. Resturants and grocers asked to avoid transgenic fish. Environmental News Network. (Internet).

Aqua Bounty Company website. 2002. WWW. aquabounty. com

Entis, E. 1997. AquaAdvantage Salmon: Issues in introduction of transgenic foods. Proc. from an international Workshop in Stockholm, May 1997, Transgenic Animals and Food Production. P. 127-131.

Fletcher, G. L., Goddard, S. V., Shears, M. A., Sutterlin, A., Hew, and C. L. 2001. Transgenic salmon: Potentials and hurdles. In: Toutant, J. P., Balazs, E. (Eds.), Molecular farming. Proc. of the OECD workshop, La Grande Motte (France), 3-6 September, 2000. INRA editions, Paris, France, pp 57-65.

Food and Agriculture Organization. 1999. The state of world fisheries and aquaculture.

Grabher C, Henrich T, Sasado T, Arenz A, Wittbrodf J, and Furutani-Seiki M, 2003. Transposon –mediated enhancer trapping in Medaka. Gene. 322: 57-66.

Hostetler, H. A., Peck S. L. and Muir, W.A. 2003. High efficiency production of germ-line transgenic Japanese medaka (*Oryzias latipes*) by electroporation with direct current –shifted radio frequency pulses. *Transgenic Res.* 12: 413-424.

Kinoshita M, Yamauchi M, Sasanuma Y, and Ozato K. 2003. A transgene and its transgenic medaka generated by the particle gun method. Zoolog. Sci. 20: 869-875.

Lu J. K., Fu B. H., Wu J. L. and Chen, T. T. 2002. Production of transgenic silver sea bream (*Sparus sarba*) by different gene transfer methods. Mar. Bitechnol. 4: 328-337.

Ninawe, A. S. 2006. Advances in aquaculture and marine biotechnology in India. *Fishing Chimes.* 26(1): 164-168, 218.

Tanaka, M. and Kinoshita, M. 2001. Recent progress in the generation of transgenic medaka (*Oryzias latipes*). *Zoology. Sci.* 18: 615-622.

Chapter 30
Techniques in Fishery Science

Collection and Preservation of Fish

Fish may be collected by suitable fishing gear. After collection, each specimen is cleaned with freshwater and is labelled, indicating its serial number, site and date of collection, gear used, colour of fins and body, presence of spots, blotches and bands etc. Scales are taken to determine the fish age. Well-galvanized or heavy tinned cans or plastic containers are used to collect fish.

The specimens should be fixed in 8-10 per cent formaline. Live fishes are to be killed in 8-10 percent formaline as they dies in this solution with all the fins expanded. Smaller specimens may directly be put in formaline solution, while medium sized prior to the fixation be given a longitudinal incision along the abdomen, without injuring the alimentary canal. Large specimens be injected 10 per cent formaline into the muscle and the abdomen, where the abdomen is not rounded but with a keel, the incision will preferably be made on the left side of the fish.

Preserved fishes should be packed in plastic jars, soaked with formaline-wet cotton, with the tail pointing upwards to avoid damage to the caudal fin and then should be transported to the laboratory for identification.

After identification fishes may be kept in glass jars or plastic jars with proper labeling in five per cent formaline with the tail pointing upwards.

Identification of Fish

Descriptive Characters

These include the morphological characters (Figures 30.1 and 30.2) of body size.

1. Body shape (cylindrical, laterally or dorsoventrally compressed)
2. Body profile (curve from the tip of the snout to the base of the caudal fin both along dorsal and ventral surface).
3. Colour of body and fins.
4. Presence or absence of barbells.
5. Presence or absence of blotches, spots or bands on body and/or fins.
6. Presence or absence of scales.
7. Lateral line complete or incomplete.

Morphometric Characters

Morphometric measurements are standard measurements that can be taken on a fish such as standard length, snout length, length

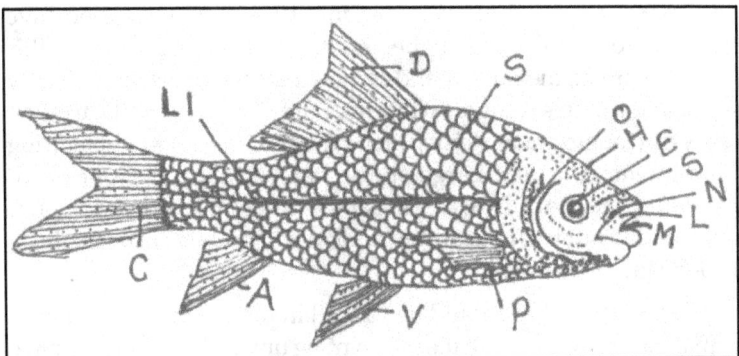

Figure 30.1: Showing Morphology of Carp Fish

H: Head; O: operculum; E: eyes; M: mouth; L: lips; Sn: snout; N: nostril; S: scales; D: dorsal fin; P: pectoral fin; V: ventral (pelvic) fin; A: Anal fin; C: caudal fin; L. I.: lateral line scales.

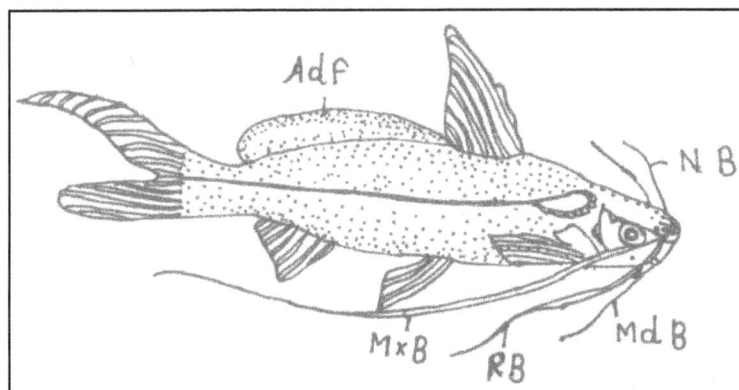

Figure 30.2: Morphology of Catfish
Adf: Adipose fin; L. I.: Latral line scales; NB: Nasal barbel;
RB: Rostral barbel; MxB: Maxillary barbell; MdB: Mandibular
barbel

of largest fin ray of the dorsal fin, depth of caudal peduncle and so on. Since these measurements change as the fish grows, these are usually expressed as ratios to standard length. Such ratios are only useful if comparisons are made between samples of fish of approximately the same size and sex, since the fish growth is not always proportional in all directions and sexual dimorphism is also noticed among fishes. Thus the morphometric measurements while vital for describing fish species may be of limited usefulness.

Length of fish is best measured by using measuring board, whereas, body parts are measured by using caliper or divider. Following are the morphometric characters of fish (Figure 30.3).

Length Measurements

1. *Total length*: The distance from tip of the mouth to posterior most end of the caudal fin is known as total length.

2. *Standard length*: It is the distance from tip of mouth to base of the caudal fin.

3. *Head length*: It is the distance from tip of mouth to posterior most edge of the opercular bone.

4. *Pre-orbital length or snout length*: It is the distance from tip of mouth to anterior margin of the orbit.

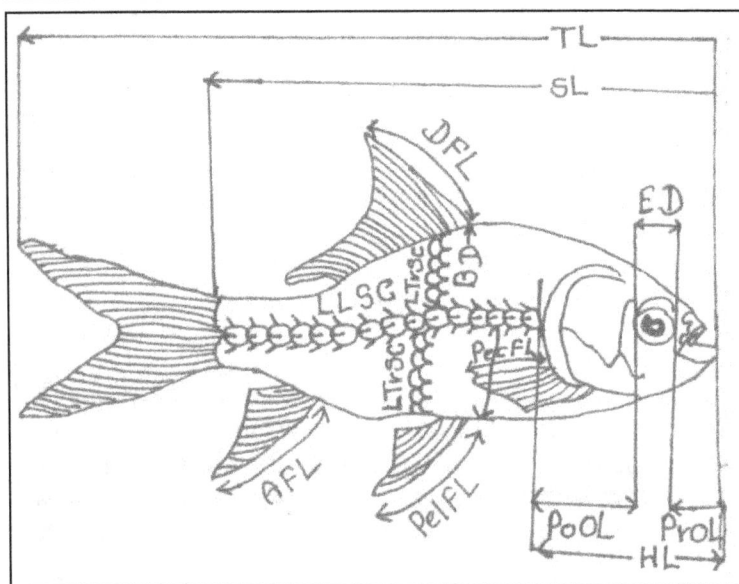

Figure 30.3: Various Morphometric and Meristic Characters of Fish

TL: Total length; SL: Standard length; HL: Head length; PrOL: Pre orbital length; PoOL+Post orbital length; DFL: Dorsal fin length; AFL: Anal fin length; PecFL: Pectoral Fin length; PelFL: Pelvic fin length; BD: Body depth; ED: Eye diameter; L. L. Sc.: Lateral line scales; L. Tr. Sc. Lateral transverse scales.

5. *Post-orbital length*: It is the distance from posterior margin of orbit to the posterior edge of the opercular bone.

6. *Pre-orbital length*: It is the distance from tip of mouth to the base of the first dorsal fin ray.

7. *Dorsal and anal fin length*: It is the distance between the anterior and posterior end of dorsal and anal fin taken along the base of fin.

8. *Pectoral and pelvic fin length*: It is measured along the longest ray of pectoral and pelvic fin.

Depth and Width Measurements

1. *Body depth*: It is the distance between the dorsal and ventral surface of the body at the deepest point.

2. *Head width*: It is the greatest distance of width of the head with gill covers held in normal position.

3. *Inter-orbital width*: It is the distance between two orbits.

4. *Eye diameter*: It is the distance between the anterior and posterior margin of eye in longitudinal axis.

5. Gape or width of mouth: It is the width of the mouth opening when the mouth is kept closed.

Meristic Characters

These are generally considered to be the most reliable taxonomic characteristics because most are easy to make and reliable. It includes anything on a fish that can be counted, such as the number of vertebrae, fin rays, spines, scale rows, lateral line scales, pores, finlets, barbels, teeth and gillrakers. Since there is often considerable variation in these characteristics within species, it is important to make the counts on adequate number of individuals so that their mean, range and standard error can be determined, if the fishes involved are to be compared with other populations.

Fin Rays

Fin rays are supported by the rays. Rays may be cartilaginous of bony. Abbreviations like D, P, V, and A, C are used for dorsal fin, pectoral fin, ventral fin, anal fin and caudal fin respectively. Few fishes have two dorsal fins (D1 and D2). D2 may be rayed or adipose. The numbers of fin rays in different fins are expressed by a fin formula. Fin rays are of two types *i.e.*, spiny rays or spines and soft rays. The spines are stiff, unbranched and unpaired, whereas, soft rays are flexible and often branched. The number of spines and soft rays are expressed by Arabic numbers *e.g.* D3/18 means, dorsal fin has three spines and eighteen fin rays.

Scales

The numbers of scales are counted along the lateral line and also in the transverse line above (starts from origin of the dorsal fin and run downward and backward to meet the lateral line) and below (starts from the origin of the anal fin and run upward and forward to reach the lateral line) the lateral line. These are represented as L. I.and L. tr., respectively. *e.g.* if we write L. I.70 and L. tr.6/8, it means that there are seventy scales on lateral line system and six scales above and eight below the lateral line.

Barbels

These are the sensory filaments, which are present near the mouth of few fishes. Number of barbells varies from 1-4 pairs. Depending upon their position, they are named as nasal barbells (nostril region), rostral barbells (snout region), maxillary barbels (on upper jaw) and mandibular barbells (on lower jaw).

Teeth

Number of pointed teeth are present on different bones like jaw bones (maxilla, mandible), vomer, palate and pharyngeal bones.

Gillrakers

These are hard and pointed structures of gill arch projecting into the pharyngeal cavity and are arranged in one or two rows. They serve to prevent the food from escaping out along with the respiratory water current.

Identification of Freshwater Culturable Fishes

1. Body covered with scales and jaws are without teeth 2

   Body devoid of scales and jaws with teeth 10

2. Scales small, silvery in colour 3

   Scales large 5

3. Anal fin long and continuous with caudal fin 4

   Anal fin short and not continuous with caudal
   fin .... *H. molitrix*

4. Dorsal profile convex .... *Notopterus notopterus*

   Dorsal profile not convex .... *N. notopterus*

5. Head covered with large plate like scales .... *Channa sp.*

   Head is not covered with large scales 6

6. Mouth wide, upturned and superior with prominent
   lower lip 7

   Mouth small and inferior or terminal 8

7. Barbels absent .... *Catla catla*

   One pair of barbells present .... *Cyprinus carpio*

8. Lips thick and fringed .... *Labeo rohita*

   Lips thin 9

9. One pair of small barbells present .... *Cirrhina mrigala*

   Barbels absent .... *Ctenopharyngodon idella*

10. Dorsal fin with spine, adipose dorsal present
    or absent .... *Mystus sp.*

    Dorsal fin spineless, adipose dorsal absent 11

11. 2 pair of barbells present and dorsal fin with 4-5 rays 12

    4 pairs of barbells present and dorsal fin with
    6-8 rays .... *H. fossilis*

12. Gape of mouth small, ends before the anterior
    margin of eye .... *Ompak sp.*

    Gape of mouth large, ends beyond the
    posterior margin of eye .... *Wallago attu*

Age Determination in Fishes

The age of a fish can be assessed from the growth rings, circuli or annuli present in the skeletal parts like the scales, vertebrae, opercular bone, otoliths, fin rays etc.

Body Scales

The scales are embedded in pockets in the skin and the embedded part is covered with striations and concentric rings. Scales to be used for age determination should be mainly plucked from the shoulder area of fish *i.e.*, between the head and the dorsal fin. The fish is first washed under tap water to remove loose scales, slime etc. The scales are ten cleaned by dipping in freshwater and rubbing each between the thumb and forefinger to remove dirt or mucus. If need be as soon as the scales are plucked from the fish, they may be washed in dilute potassium hydroxide solution before mounting. The scales should be mounted with their convex side up on microscopic slides. While observing the scales under a microscope, the field of the microscope should appear dark except for a small segment at the bottom, which should be very bright. Scales should be observed under different magnifications so as to have a clear view of the scale structure. Ordinarily, the scales will adhere directly to the slide once they have dried. However, for some scales a drop of egg albumen, glycerin or gelatin containing fungicide or bactericide is used for mounting. It is also advisable to take two scales one from each side of the fish, to serve as a check against each other. A sample of five scales is sufficient for the age assessment.

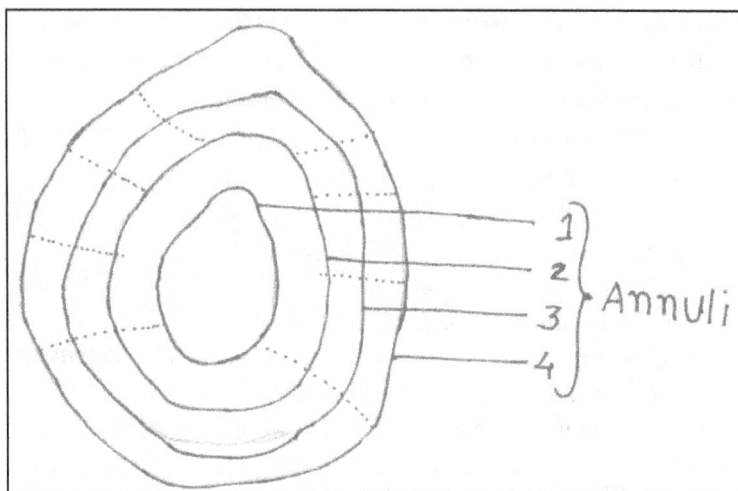

Figure 30.4: Body Scale with Annuli Shown

The structure discontinuities used for age determination result from irregularities in the pattern of the sclerites, which may be slightly distored or closely spaced. When the discontinuities are narrow, they are usually termed 'rings' or 'annuli' as these are formed yearly. The center of the scale is known as 'focus'. The number of rings or annuli in a scale depends on the nature of the fish growth. The scales mounted in slides should be labeled and stored for future reference.

Otoliths

Three otoliths are found in the sacculus of the inner ear on either side of the head. Among them, the largest otolith, viz; the sagitta is mainly used for age determination. To collect the sagitte, the fish are first boiled in 2-5 per cent potassium hydroxide solution. The head bone is then dissected and the otoliths are removed using a pair of fine tweezers and are preserved dry in paper envelopes. Otoliths can also be preserved in a mixture of ethyl alcohol and glycerine (9:1). To brighten the otoliths and to accentuate the appearance of growth rings, glycerine, xylol, or cresol may be used.

Otoliths are first cleaned and polished and then observed under a microscope. For larger otoliths, cross-sections are obtained through

Figure 30.5: Position of Otolith

the nucleus using a fine knife. A graphite stone or grinding and polishing machines are often employed for polishing the otoliths. As in the case of fish scales, the growth rings or annuli present in the otoliths are counted and the age of the fish is assessed. This method is more reliable than the scale method because of the formation of thick annuli in otoliths.

Opercular Bones, Spines and Vertebrae

While the opercular bones are collected by excising, the spines and vertebrae are collected by boiling the fish in 2-5 per cent potassium hydroxide solution. To observe the vertebrae, cross-sections must be taken. The year mark or annular rings of these hard parts are normally used for age determination.

Length-Frequency Distribution Method

About 2000 fishes belonging to various periods and length groups are collected random from the fish landing centers. The lengths are measured by means of a measuring board before grouping them into different size classes with an interval of 5mm or 10mm. The monthly percentage frequencies are then calculated and their growth rates are subsequently traced by plotting. Before drawing conclusion, the direct and direct methods should be compared. It is also worth mentioning that this method should not be used for continuous breeders.

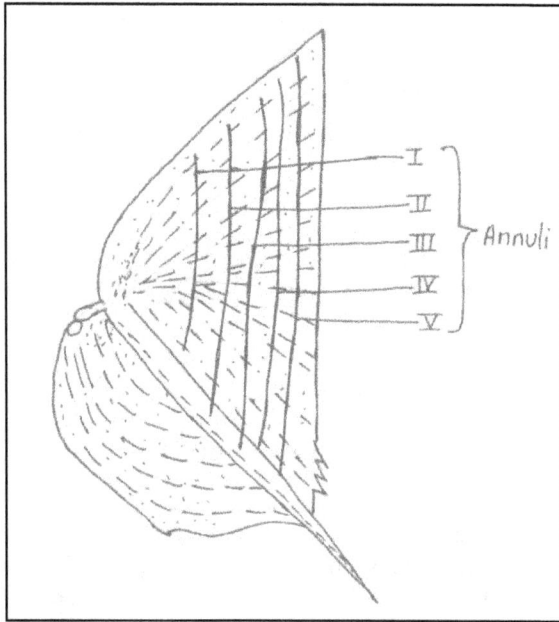

Figure 30.6: Opercular Bone with Annuli Shown

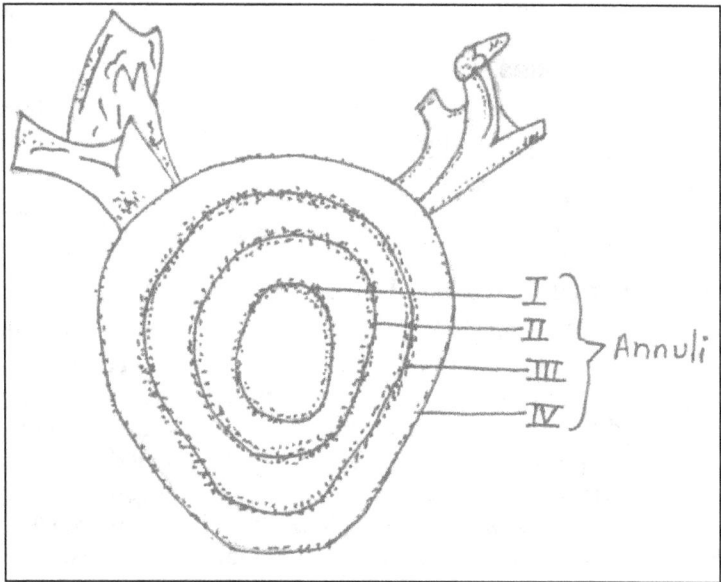

Figure 30.7: Vertebra with Annuli Shown

Marking and Recapturing Method

It is a direct and positive way of determining the age of fish. In this method, individuals of known age are marked and released in water body. They are then recaptured at intervals of time and their growth is released. The marking or tagging should not affect the health, mobility and yield of fish. This method is of limited practical value because marking or tagging of fish is costly and time-consuming operation. Moreover, the chances for recapturing the marked fish is low in large water bodies.

Growth Determination in Fishes

Growth means change in length and/or weight with an increase in age. Length-weight plotted in relation to time, produces a vector diagram, which is known as growth curve. This curve is 's' shaped which depicts growth rate or change in length or weight of fish with time.

Fish growth is determined by using the following methods:

Known Age Method

Experiments are conducted with fishes of known age, at least for one year. Fish samples are collected periodically to determine the growth.

Marking and Recapturing Method

Fishes of known length are marked and released in natural waters. These are then recaptured at intervals of time and growth is recorded.

Length Frequency Method

It is helpful in determination of average growth rate of fish. Many fishes are seasonal breeders and their population is composed of distinct stocks. Each fish of a population is measured and average length is calculated. Data is taken for 2-3 successive years during the same season. Peaks represent different age groups. Shift in peak towards larger size indicate rate of growth. This method is applicable to those fishes, which have long span of life. It does not hold good to the older age groups in which growth rate becomes slow. Further it is not useful in fishes, which are non-seasonal breeders.

Body-scale Relationship Method

Body lengths for each age are back calculated from the body length and total scale radius relationship and measurement of radius of annulus. This method works well with fishes showing a linear and directly proportional body-scale relationship. Following formula is used to calculate the fish length and growth

$$L_n = \frac{L \times S_n}{S}$$

where,

L_n: Length of fish at the time of annulus 'n' formation

L Length of fish when scale sample was taken

S_n Radius of annulus 'n'

S: Total scale radius.

Ova Diameter

It is used to know the size of ova present in ovaries. It is studied from the preserved ovaries. A small portion of ovary from the middle region to be teased on a slide to separate the ova. A random sample of 200 ova is taken and their diameter is measured with the help of ocular micrometer keeping the magnification constant and represented graphically. Ova diameter serves as a key to maturity stages and it is also an indicator of maturity, spawning season and spawning periodicity.

Fecundity of Fishes

Studies on fish fecundity are receiving much attention now days. It plays a key role in fish stock management. Fecundity is a measure of the reproductive capacity of a female fish, and is an adaptation to various conditions of the environment.

Methods of Estimation

Gravimetric Method

After their liberation from the ovarian tissues, the ova are thoroughly washed and spread on a blotting paper to dry in air. The total number of ova so collected are then weighed and random samples of about 500 are counted and weighed. The total number of ova in the ovaries is then obtained from the equation given below:

$$F = \frac{nG}{g}$$

where,

    F: Fecundity

    n: Number of ova in the subsamples

    G: Total weight of the ova; and

    g: Weight of the subsample in the same unit

Volumetric Method

The ova are transformed to a measuring cylinder containing a known volume of water and the total volume is recorded. Subsamples are then taken by shaking the container, pipetting out a subsample of known volume and counting the number of ova in the sample. Fecundity is then estimated by using the following formula:

$$F = \frac{nV}{v}$$

where,

    F: Fecundity

    n: Number of ova in the subsample

    V: Volume to which the total number of ova is made; and

    v: Volume of subsample.

Von Bayer Method

By studying the volumetric and gravimetric methods Von Bayer prepared a chart from where by knowing the average diameter of the egg, it is easy to calculate the total number of eggs present per quart of liquid.

Firstly find the average diameter of the eggs by means of a small metal trough graduated in tenths of inches, remove excess moisture from between the eggs. Make at least three determinations based on as. Many eggs as ruler in through will permit, and obtain average diameter. Then consult the Von Bayer table giving the number per quart for eggs of various diameters.

Best Method for Fecundity Estimation

Volumetric or gravimetric method gives more accurate results than Von Bayer method.

The most accurate enumeration of fish eggs is probably by actual count but this can be very tredious and time consuming. When actual count of eggs impracticable approximate number may obtain by method described below; for that we have to complete the following table using routines given below for each method. If possible repeat each method several times and can calculate the results statistically.

Table 30.1: Comparison of Results of Various Methods for Determining Number of the Eggs

Method	Estimated No. of Eggs (Mean)	% Error (of mean)	Fiducial Limits– Range at 5% Level
Volumetric method			
Gravimetric method			
Von Bayer method			

Estimation of RBC Present in Cubic mm Volume of Blood

Requirement

Haemocytometer, blood sample, microscope, Hayan's solution, Prickling needle.

Structure of Haemocytometer

Glass slide, center 'H' shaped chamber, 2 prominent ridges, each ridge with 5 cubical chambers, each cubical chamber further sub-divided into 25 sub-divisions.

 ☆ 4 slide cubical chambers each with 16 subdivisions

 ☆ Total division in central chamber are (25x16=400)

 ☆ Central chamber is used to count RBC

 ☆ Haemocytometer includes a pipette.

Hayan's Solution

It contains mercuric chloride (0.5g), sodium chloride (1.0g), distilled water (200ml) and sodium sulphate (5.00g).

Mercuric chloride acts as a corrosive sublimate and fixes the RBCs in the blood. The other ingredients act isotnically so that RBC's may not burst due to haemolysis. Hayan's solutions also serve to dilute the blood.

Procedure

First clean and dry mixing pipette (dilution 1:2000). Take the fish in hand and cut the tail so that blood flows freely. If necessary squeeze it. Leave the first few drops to flow away. Suck blood with RBC pipette upto the 0.5 mark slowly and carefully. Wipe the tip of the pipette. Suck Hayan's diluting fluid up to 101 mark (Hayan's fluid prevent haemolysis, coagulation and bacterial growth). Mix and shake thoroughly for a minute. Transfer the blood on to the counting chamber covered with a cover glass. Avoid the overflow of mixture from the chamber and trapping of air bubbles. Allow the cells to settle. Bring counting scale into the focus under a microscope.

Counting

Select smaller squares for counting, taking one from each four corners and fifth one from the center.

Calculation

$$\text{No. of RBC/cubic mm} = \frac{\text{No. of RBC counted} \times \text{Dilution} \times \text{Depth}}{\text{Area of chamber counted}}$$

Depth of chamber = 0.1mm

Dilution of blood = 200 times; and

Area counted is 80/400 = 1/5 sq. mm

= Number of RBC countedx200x10/1/5

= Number of RBC countedx10, 000

Determination of Haemoglobin Contents of Fish Blood

Fish is an aquatic vertebrate and its blood is liable to be affected by a variety of substances occurring in the media, therefore, measurement of haemoglobin contents will yield a lot of information about the environment, pathological conditions and physiological state of the fish. Haemoglobin is present in the erythrocytes and is a principal oxygen carrier substance.

Requirements

Haemoglobinometer and live fish.

Procedure

Fill the central diluting tube with 20 cu mm of N/10 Hcl (9ml of Hcl in 1L of distilled water). Bleed the fish and suck 20 cu mm of blood in a pipette/capillary. Transfer the blood by blowing the pipette/capillary gently in the diluting tube and rinse the capillary several times so that no trace of blood is left in it. After 10 minutes add drop by distilled water or N/10 Hcl with continuous stirring till the colour of blood in the diluting tube start matching with the standard tubes. Then record the dilution of blood and also note the point where its colour matches with the colour of the standard tube and calculate haemoglobin as mg ml^{-1} of blood.

Food and Feeding Analysis of Fish

Digestive tract is taken out from the dissected fish and its length is measured and both the ends are tied. It is then preserved in five per cent formaline.

Relative Gut Content

The gut length of fish depends on the nature of food they consume. The length increases with increasing proportion of vegetative matter in the diet. Its length varies from fish to fish at different stages of its life. Relative gut length is calculated as:

$$\text{Relative gut length} = \frac{\text{Length of gut}}{\text{Total body length of fish}}$$

If relative gut length is less than unity, then fish is carnivorous and if more, the fish is herbivorous. An intermediate value indicate that the fish is omnivorous.

Feeding Intensity

The degree of feeding is known as feeding intensity. It can be ascertained by examining fullness of stomach and by calculating the Gastrosomatic Index (GasI) by the following formula:

$$\text{GasI} = \frac{\text{Weight of stomach content}}{\text{Weight of fish}} \times 100$$

GasI of several fishes show seasonal variations in both the sexes and is maximum during post spawning period and minimum during the breeding season. The feeding intensity of fish varies with the season, state of maturity, spawning season and the availability of food items.

Gut Content Analysis

Quantitative and qualitative analysis of gut contents is done by the following methods:

Numerical Method

The total number of individuals of each food item is recorded and expressed as percentage of the total number of food organisms in the stomachs examined.

Frequency or Occurrence Method

Different types of food items are sorted out and record is made of the presence or absence of different food items and number of fishes (stomach in which a certain item is present). Frequency of occurrence of an item is determined by the number of stomachs showing the presence of that item. It is expressed as the percentage of the total number of fish stomach analysed. This method gives the frequency of particular food item along with qualitative analysis.

Volumetric Method

It presents the relative importance of each food item by volume. First total volume of the entire stomach content is determined by water displacement. The different food items present are segregated and volume of each item is obtained in the same way. The volumetric values of different items are expressed as per cent of the total volume of entire stomach contents. This method does not take into account the number or frequency of occurrence of a given food item. Hence, when used in combination with numerical method, it gives a better picture of the feeding habit than when it is used exclusively.

Gravimetric Method

The total dry weight of the stomach content is taken. The different food items are segregated and the weight of each item is taken. The weight of different food items are expressed as percentage of the total weight of stomach content.

Classification of Fish Based on Feeding Habits

Depending Upon the Food Item

Herbivorous Fishes

Fishes which feed on filamentous algae, unicellular algae, portion of higher aquatic plants along with some sand or mud. Their 75 per cent or more of the gut contents plant material, hence are known as herbivorous fishes *e.g.*, *Labeo sp*, *Ctenopharyngodon idella*, *Tilapia mossambica*, *Schizothorax spp* etc.

Omnivorous Fishes

These fishes consume food of all kinds *i.e.*, plants, animals, debris and mud *e.g.*, *Cirrhina mrigala*, *P. ticto*, *P. sarana*, *Clarias batrachus*, *Heteropneustes fossilis*, *Cyprinus carpio* etc.

Carnivorous Fishes

The fishes feed mainly on animals like crustaceans, insects. larvae, molluscs, tadpoles, smaller fishes etc. *e.g. Channa sp*, *Wallago attu*, *Rita rita*, *Mytus sp* etc.

Plankton Feeders

Some fishes feed on phyto and zoo-plankton, which they obtain by filtering water through their gill-raker e.g. *Hilsa ilisha*, *Catla catla*, *Hypophthalmichthys molitrix* etc. They are both omnivorous and carnivorous.

Debris Feeder

These fishes feed mainly on mud and detritus *e.g.*, *Xenocypris macrolepis*.

Depending Upon the Niche

Surface Feeder

These fishes take food from the surface of water *e.g.*, *Catla catla*, *Ambassis nama*, *A. ranga*, *Hilsa ilisha*, *Hypophthalmichthys molitrix* etc.

Column Feeder

These fishes are neither true surface feeders, nor true bottom feeders. They mostly depend on the organisms of the mid water. Such species are known as column feeder *e.g.*, *Wallago attu*, *Tor tor*, *Labeo rohita* etc.

Bottom Feeder

These fishes take food from bottom of the water body *e.g.*, *Clarias batrachus*, *Heteropneustes fossilis*, *Cirrhina mirigala*, *Channa marulius*, *Channa striatus* etc.

Aggregation and Dispersion of Chromatophores in Fishes

Fishes are brightly and brilliantly coloured. Pigments of different colours are usually present in the skin, in cells called chromatophores. These cells commonly contain a black or brown pigment called melanine and are known as melanocytes. Yellow (xanthine) and red (carotene) pigments also occur in the skin of vertebrates and are contained in chromatophores called xanthophores and erythrophores respectively. Some chromatophore containing pteridine platelets, which give an iridescent appearance, are known as iriodophores.

Many fishes are capable of changing their body colour, this may be of semi-permanent nature which is slow process or a temporary change which is fairly rapid process. A rapid adjustment in colour is brought about by rearrangement of pigment granules. Due to aggregation of pigments when they move towards the center of cell the colour of skin becomes pale, and their dispersion when they move towards the periphery, the skin colour becomes dark.

There are two mechanisms to control and coordinate changes in the chromatophores of the body. This is either through the hormones or through the nervous system.

Procedure

First select three fishes belonging to same species. Then keep them in transparent glass jars. Out of them one species is treated as normal. One fish is injected with adrenaline and another will be injected with atropine sulphate. These injections should be administered at the base of pectoral fin or above the lateral line at the caudal region. Then after sometime observe the injected two fishes, and compare their colour changes with the normal fish and draw the conclusion.

Result

Adrenaline injected fish becomes pale in colour as compared to the normal fish. It is due to aggregation of the chromatophore pigments. Atropine sulphate injected fish becomes darker in colour as compared with the normal fish. It is due to the dispersion of chromatophore pigment.

Types of Scales in Fishes

In most of the fishes exoskeleton is present in the form of scales. Only few fishes are naked. They have no scales on body.

Cosmoid Scales

These are found in extinct crossopterygii and dipnoi. External layer of scale is thin and enamel like. It is known as vitrodentine. The middle layer is made up of hard, non cellular and dentine like material called the cosmine. Cosmine contains a large number of branching tubules and chambers. The inner layer is made up of bony substance called isopedine. These scales grew by the addition of new isopedine material from below, along the edges. Cosmoid scales are not found in the living fishes. These scales are regarded as the precursor of the ganoid, placoid and the bony scales of modern teleosts.

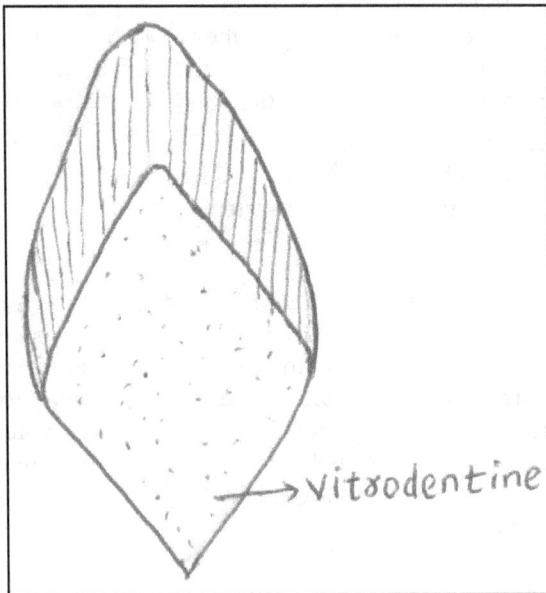

Figure 30.8: Cosmoid Scale

Ganoid Scales

These are characteristic of primitive actionpterygians called ganoid fishes. These are of various forms and structure. Scales are heavy and have an outer layer of hard inorganic, enamel like material called ganoine. The middle layer is cosmine, which contains the numerous branching tubules. The innermost layer is thickest, which is made up of lamellar bone called isopedine. These scales grow by the addition of new layers to lower as well as upper surface. These scales are rhomboid in shape and articulate by peg and socket joints. These scales are best developed in chondrostean and holostean fishes.

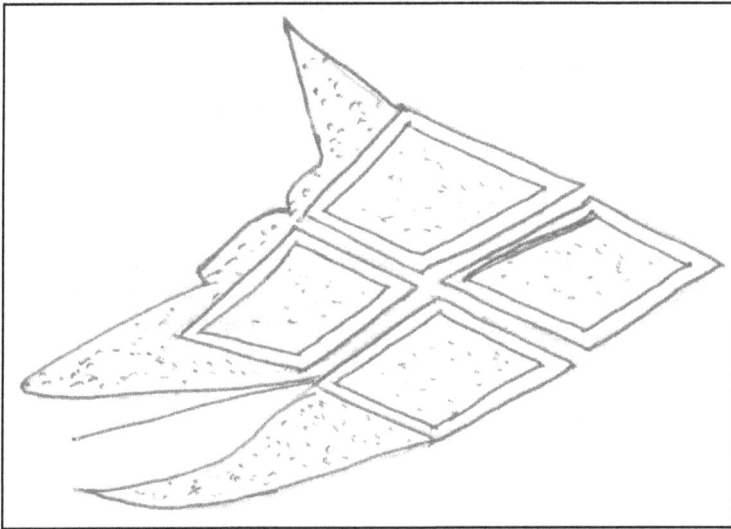

Figure 30.9: Scale of Lepidosteus

Placoid Scale

Placoid scales, found in elasmobranch are very similar to the denticles of ostracoderms. They have dentine laid down by osteocyte cells and the pulp cavity, ramified in a manner similar to the teeth of vertebrates. Each scale has a disc-like basal plate embedded in the dermis and a spine projecting out through the epidermis. The spine has an external covering off enamel like, hard, transparent material called vitrodetine. This is followed by a layer of dentine enclosing a

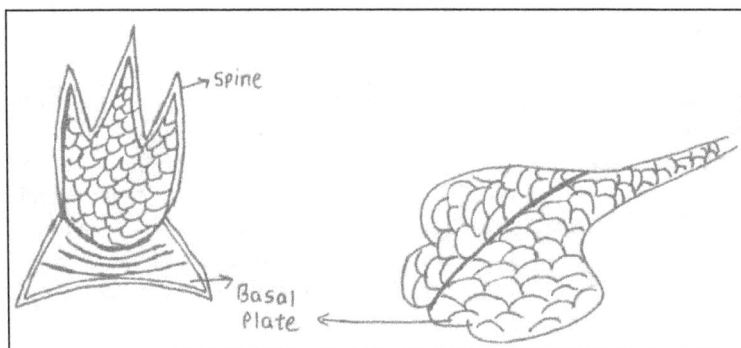

Figure 30.10: Placoid Scale

pulp cavity from which several branching dentinal tubules radiate in different directions. The center of the basal plate is perforated by an aperture to provide entrance to the blood vessels and nerves from the dermis.

Bony Ridge Scales

The cycloid and ctenoid scales are also known as the scales of modern fishes as well as the Bony Ridge Scales. They are present in majority of the teleostean fishes. These are flexible and transparent structures. These scales exhibit characteristic ridges alternating with

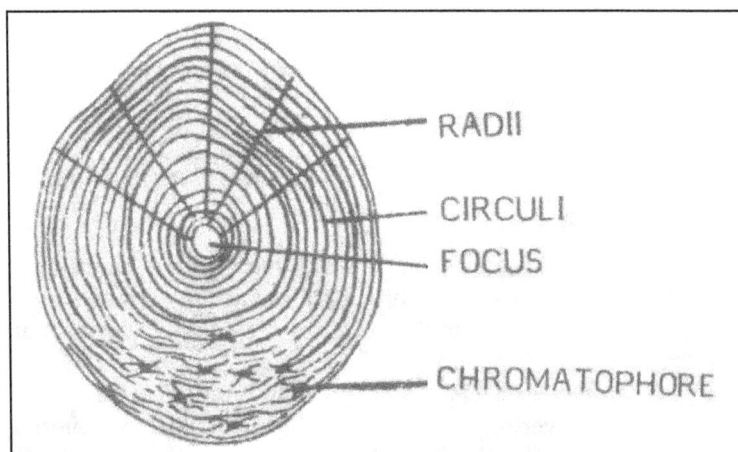

Figure 30.11: Cycloid Scale

grooves and generally the ridges are in the form of concentric rings. The central part of the scale is known as the 'focus'. Focus is the first part to develop. In many species, oblique grooves or radii run from the focus towards the margin of the scale.

Cycloid scales are thin and roughly rounded in shape, being thicker in the center and thinning out towards the margin. They form protective covering over the skin and project diagonally in a imbricating pattern. The part of the scale which is exposed to view in situ condition (posterior area) generally shows less distinct ridges or circuli and chromatophores are also sometimes attached to it. The anterior area lies embedded in the skin. They are found in number of teleostean fishes having soft rayed fins *e.g.*, *Labeo rohita, Catla catla, Cirrhina mrigala* etc.

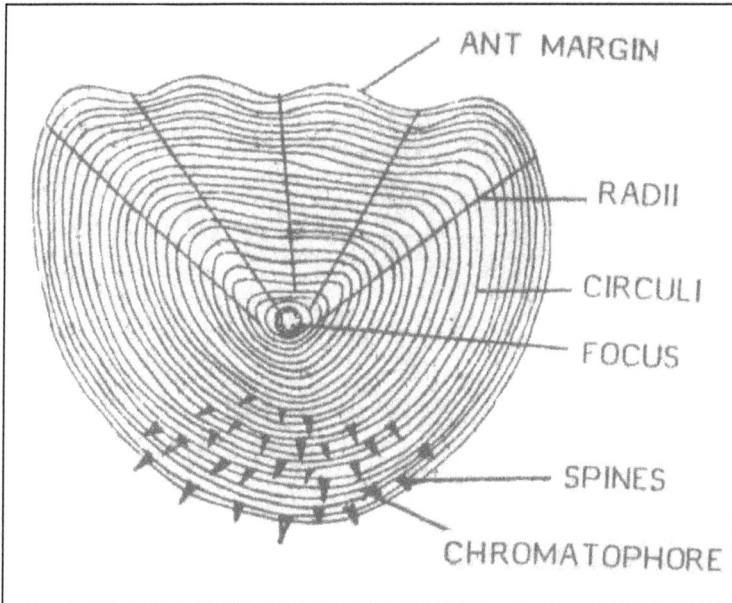

Figure 30.12: Ctenoid Scale

Ctenoid scales are circular and can be differentiated from the cycloid by having more or less serrated free edge. Several spines are present on the surface of the posterior area of the scale. These scales are found in large number of fishes with spiny-rayed fins.

Index